THINGS
COME
TO LIFE

THINGS COME TO LIFE

Spontaneous Generation Revisited

HENRY HARRIS

OXFORD

UNIVERSITY PRESS

Great Clarendon Street, Oxford OX2 6DP

Oxford University Press is a department of the University of Oxford.
If furthers the University's objective of excellence in research, scholarship,
and education by publishing worldwide in

Oxford New York

Auckland Bangkok Buenos Aires Cape Town Chennai
Dar es Salaam Delhi Hong Kong Istanbul Karachi Kolkata
Kuala Lumpur Madrid Melbourne Mexico City Mumbai Nairobi
São Paulo Shanghai Singapore Taipei Tokyo Toronto
and an associated company in Berlin

Oxford is a registered trade mark of Oxford University Press
in the UK and in certain other countries

Published in the United States
by Oxford University Press Inc., New York

First published 2002

A catalogue record for this title is available from the British history

Library of Congress Cataloging in Publication Data

Harris, Henry, 1925–
    Things come to life : spontaneous generation revisited / Henry Harris.
    1. Spontaneous generation.   2. Life—Origin.   I. Title.
QH325 .H35 2002   576.8′3—dc21   2001054856

ISBN 0 19 851538 3(Hbk)

10 9 8 7 6 5 4 3 2 1

Typeset by Integra Software Services Pvt. Ltd., Pondicherry, India
www.integra-india.com

Printed in great Britain by
on acid-free paper by T. J. International Ltd, Padstow

How could hair come from what is not hair
or flesh from what is not flesh?

Anaxagoras

Preface

Old ideas die hard. The *De revolutionibus orbium coelestium*, in which Copernicus proposed that the earth revolved around the sun, was published in 1543. As a statement of fact, it was condemned by the Holy Office of the Catholic Church in 1616 and remained on the index of prohibited books until 1835. Harvey's *De motu cordis*, in which he demonstrated the circulation of the blood, was published in 1628. But for a generation or more after that there were many eminent scholars across Europe who refused to accept his findings. Newton's *Opticks*, the definitive version of his observations on the composite nature of white light, appeared in 1704. But his conclusions were assailed more than a century later by no less a figure than Goethe. And Darwin's *Origin of species*, which appeared in 1859, still has any number of intransigent opponents.

Spontaneous generation—the theory that, under appropriate conditions, inanimate matter can generate living beings by completely natural processes—is an idea that has proved to be even more difficult to dislodge. It has its origins in the mists of time, and although some classical authors had reservations about it, it was not put to any experimental test until the seventeenth century. Thereafter it remained the subject of debate, often acrimonious, until the beginning of the twentieth. This debate of course mirrored the theological and philosophical preconceptions of those that engaged in it, but spontaneous generation was not disposed of by means of argument. It was whittled away by a long series of increasingly sophisticated experiments. Not that there was ever an *experimentum crucis* that settled the matter once and for all; there is no such thing as a perfect experiment. It was rather that the idea simply ceased to be of interest because the best experimental evidence consistently failed to support it.

This book is, in large part, a commentary on the experimental evidence. The personal, social, and religious backgrounds of the protagonists have not been underestimated; but, in my view, these contextual elements have had less influence on the outcome of the debate than some authors have proposed. What, in any case, has not so far been seriously discussed is the probative value of the experiments themselves. This I have now attempted to do. But I have laboured to avoid losing the main lines of the experiment in a welter of technical detail; and what I have written will, I hope, be easily understood by those who read these pages, whether they have a scientific background or not. The history of science will not yield a more than superficial view of the growth of human knowledge unless it deals not only with the scientists but also, and no less important, with the science itself.

I thank Professor Herman Waldmann of the Sir William Dunn School of Pathology for his hospitality and Professor Peter Cook for his help with the diagrams.

Henry Harris
Oxford
Hilary 2000

Contents

CHAPTER 1

Beliefs

In a world peopled by gods with magical powers, the creation of a living organism from inanimate matter presented no particular conceptual problem. Virtually all ancient religions assert that, in the beginning, all life arose from the earth, or from the sea, or, in any case, from materials that were initially 'without form and void'. This act of generation is usually seen as the result of some kind of divine intervention, and although the timescale of the process is only rarely specified, ancient faiths all describe a more or less abrupt event, or at least one that took place within an easily measurable period of time. So long as it was held that the generation of life from inanimate materials had something of the miraculous about it, the idea that this could at any time be repeated was little more than an expression of continued belief in the possibility of miracles. But it was not an inevitable consequence of that belief, for it could still be argued that whatever divinity was responsible for the initial creation chose not to repeat it; and that argument in various forms was in fact a widely held theological orthodoxy for centuries. However, ancient texts take into account the commonplace observation that the living forms originally created from inanimate material by divine intervention are so created that they thereafter reproduce themselves by the transmission of 'seed'. In Genesis from the Old Testament, for example, we read: 'The earth brought forth grass and herb yielding seed after its kind.' It was, of course, well known as far back as historical records go that some plants could be propagated by other, non-magical, procedures. But there remained a large enough group that could not be so propagated and that produced no visible seed. Similarly, there were many animal species in which transmission of 'seed' from the male to the female could not be detected. How such animals and plants were generated remained obscure, and it was this obscurity that engendered yet another belief: that, under certain circumstances, inanimate matter could generate life 'spontaneously'.

Spontaneous generation is a theory that requires some elaboration. It proposes that apart from the miracles that require divine intervention, living forms can also arise within inanimate material by a completely natural process, one that recurs whenever conditions are right. In his *Historia animalium*[1] Aristotle (384–322 BC) makes a clear distinction between generation via 'seed' and 'spontaneous' generation, and it is clear that he regards both as established facts. In a passage quoted by William Harvey (1578–1651) at the beginning of Section Two of his *De generatione animalium*[2] we read:

Animals have one thing in common with plants, that some arise from seed and others spontaneously; for plants arise from the seeds of other plants or else spring up spontaneously, having acquired some principle capable of producing them, and some of these attract nourishment to themselves from the ground while others are usually bred in other plants, so some animals are generated from parents of their kind while others arise spontaneously and not from any antecedent germ or seed that is kin to them. Some of these are generated from putrefying earth or plants, as is the case with a number of insects or are engendered inside the animals and out of the excrements of their parts.

To convey the idea of spontaneity, Aristotle and Theophrastus (*c*.370–*c*.288 BC), his successor at the Lyceum, use the word αυτοματον, which can be literally translated as 'acting of its own accord'; and Harvey uses the classical Latin word *sponte*, which, in this context, means much the same thing. From *sponte* we have spontaneous generation, *génération spontanée*, *generazione spontanea*, but we have no cognate equivalent in German. German biologists continued to use Latin terms until the later years of the nineteenth century: *generatio spontanea*, *generatio primitiva*, but most commonly *generatio aequivoca*, which gave expression to the uncertainty of the whole process.

Reservations about spontaneous generation can already be found in Theophrastus. Although he accepts Aristotle's proposition that living forms can be generated in this way, he has some doubts about the evidence in certain cases. Thus, in his *Historia plantarum*[3] he writes:

> The ways in which trees and plants in general originate are these: spontaneous growth, growth from seed, from a root, from a piece torn off, from a branch or twig, from the trunk itself... Of these methods spontaneous growth comes first, one may say, but growth from seed or root would seem most natural... Spontaneous generation, broadly speaking, takes place in smaller plants, especially in those that are annuals and herbaceous. But still it occasionally occurs too in larger plants whenever there is rainy weather or some peculiar condition of air or soil. Many believe that animals also come into being in the same way.

But this last sentence also indicates that, even at that time, there were some who did not believe it. Doubt is more directly expressed in the following passage:

> Of the sterile sorts, one might, rather, expect them to be spontaneous, as they are neither planted nor grown from seed, and if they come to be in neither way, they must necessarily be spontaneous. But this may possibly not be true, at least for the larger plants; it may be, rather, that all the stages of the development of their seeds escape our observation... Indeed, the development of seed escapes observation also in many of the smaller herbaceous plants... Further, in trees too, some seeds are hard to see and small in size...

Theophrastus, echoing Aristotle, calls for more investigation of the question:

> Let this be given merely as our opinion; more accurate investigation must be made of the subject and the matter of spontaneous generation must be thoroughly inquired into. To sum the matter up generally: this phenomenon necessarily occurs when the earth is thoroughly warmed and when the collected mixture is warmed by the sun, as we see also in the case of animals.

In the matter of the temperature at which the reaction occurs, Theophrastus disagrees with Aristotle, who believed that all forms of spontaneous generation were associated with decomposition. Theophrastus, however, states that 'nothing is produced by decomposition, but instead by undergoing concoction'. Spontaneous generation was, in his view, a process that required heat.

Some three hundred years later Lucretius (94–55 BC) was still worrying about the plausibility of spontaneous generation. The famous phrase in the first part of the *De rerum natura*,[4] asserting that nothing can be generated or created out of nothing, may well have a broader significance than that implied by spontaneous generation, but there is little doubt that Lucretius also had living forms in mind. 'For if they could be made out of nothing,' he writes, 'every kind of living creature could be generated out of anything, and none of them would require seed.' H. J. Abrahams quotes a passage from the *De rerum natura* which he interprets as evidence for the view that Lucretius did indeed believe in spontaneous generation.[5] But this is certainly not the case for spontaneous generation as I have defined it. Here is the relevant passage: 'For neither can animals fall from the sky, nor living things be produced by briny pools. The only remaining conclusion is that the earth is deservedly given the name mother, seeing that all things are created from the earth.'[6] The first sentence is clearly a rebuttal of fanciful theories advanced by ancient writers, both Greek and Roman, about how living forms are, or were, originally generated. The second sentence reluctantly admits the conclusion, reached not by observation but by elimination, that, after all, the earth must be the origin of all things. This, it seems to me, does not assert that inanimate soil generates living creatures, but is merely an expression of the conventional attitude that the earth is the original mother of everything.

Be that as it may, belief in spontaneous generation and scepticism about its existence co-existed in the Graeco-Roman world and for centuries afterwards. Indeed, the idea that inanimate matter could generate living things without divine intervention was not finally put to rest until the end of the nineteenth century. For example, in the *Metamorphoses*, Ovid (43 BC–AD 17) contends that frogs and even parts of frogs are generated from mud; and, accepting this, Pliny (AD 23–79) states that, after a few months of life, these frogs turn back into mud. Indeed, according to Ovid, dank vapour can create anything, and Pliny specifically mentions that many flying insects are produced by the dank dust of caves. Lucian (AD c.117–c.180) reports that flies are generated by human corpses. Several authors agree that wasps are generated from the dead meat of horses and bees from the putrid viscera of animals, especially bulls. Celsus (AD 14–37) disbelieves that bees can be generated everywhere, or there would be no point in looking for them in the viscera of bulls. Pomponius Mela (fl. AD 40) provides perhaps the most dramatic statement of the spontaneous generation hypothesis. In discussing the waters of the Nile, he writes that they are efficient producers of all sorts of living things: fish, crocodiles, even huge wild animals. Moreover, the fluvial waters are said to fructify the soil and confer on it similar generative powers.

A modern reader, fortified by two millennia of observation and experiment, would view these propositions with condescension and perhaps amusement, but it remains a remarkable fact that the belief that living creatures, or at least some living creatures, could be formed spontaneously from inanimate matter survived the transition from paganism to Christianity, from polytheism to monotheism, from scholasticism to empiricism. Even in the seventeenth century, Johannes Baptista van Helmont (1579–1644), who straddled the worlds of alchemy and modern chemistry and who enjoyed an enormous European reputation in his time, believed that frogs, slugs, and leeches were generated spontaneously and, indeed, offered a recipe for the spontaneous generation of mice: 'If a dirty shirt is stuffed into the mouth of a vessel containing wheat, within a few days, say 21, the ferment produced by the shirt, modified by the smell of the grain, transforms the wheat itself, encased in its husk, into mice.'[7] And William Harvey, whose De motu cordis still stands as an exemplar of experimental physiology, nonetheless states categorically that spontaneous generation ('sponte nascentia') is one of the two ways in which animals can be generated, the other, of course, being from existing parents.[8]

The durability of these beliefs was due to two related factors: the ecclesiastical scholasticism of the Middle Ages and the intellectual dominance of Aristotle. Learning, for more than a millennium, was very largely in the hands of the Church, which did not encourage, indeed militated against, empirical enquiry. And at the centre of the learning sanctioned by Church doctrine were the works of Aristotle. It was not until the scientific revolution of the seventeenth century that assertions made by Aristotle were systematically challenged. Even those who, like van Helmont and Harvey, made major contributions to the development of empirical enquiry in other contexts saw nothing questionable in Aristotle's views on spontaneous generation.

The passage of the centuries did nonetheless produce a progressive narrowing of the range of living things that were still thought to be generated spontaneously. With the continual growth of knowledge concerning the modes of reproduction of both animals and plants, and in the absence of any credible evidence that larger species could arise fully formed from inanimate matter, the attention of men of learning turned inevitably to the study of smaller creatures or those whose mode of reproduction remained obscure. By the later years of the seventeenth century the only living things still widely believed to be generated spontaneously were insects, the smallest creatures then known. Aristotle singles out insects as an example of spontaneous generation taking place in putrefying earth or plants; and Harvey also discusses insects as an instance of life arising from inanimate matter. But the eventual restriction of belief in spontaneous generation to insects was not driven only by common experience.

The seventeenth century saw a sea change in the way in which enquiry was carried out and especially enquiry into natural phenomena. Even van Helmont, who believed in the spontaneous generation of mice, sought to overcome the baneful influence of useless logic ('entia rationis'); as early as 1619,[9] Johann Valentin Andreae, an influential protestant divine, wrote that if you did not analyse matter through experiment you were worthless; and the Royal Society of London for Improving Natural Knowledge, founded in 1660, chose as its motto the now famous *Nullius in verba*, which was a direct challenge to the authority of received texts. During the seventeenth century, learning in Europe was swept by a tide of uncompromising empiricism, but belief in the spontaneous generation of insects did not at once disappear in the tide. That had to await the publication and dissemination of Francesco Redi's great book (Fig. 1.1), to which we now turn.[10]

ESPERIENZE
Intorno alla Generazione
DEGL'INSETTI
FATTE
DA FRANCESCO REDI
Gentiluomo Aretino, e Accademico della Crusca

E da Lui scritte in una Lettera

ALL' ILLVSTRISSIMO SIGNOR
CARLO DATI.
Quinta Imprefsione.

IN FIRENZE, MDCLXXXVIII.
Nella Stamperia di Piero Matini, all' Insegna del Lion d'Oro.
CON LICENZA DE' SVPERIORI.

Fig. 1.1 Title-page of Redi's *Experiments on the generation of insects*

Notes

1. The excerpts from Aristotle are given in Gweneth Whitteridge's translation of Harvey's *De generatione animalium*. See note 2.
2. W. Harvey (1651), *De generatione animalium* (ed. and trans. Gweneth Whitteridge). Blackwell, Oxford (1987).
3. The excerpts from Theophrastus are given in the translation by G. E. R. Lloyd in (1987), *The Revolutions of wisdom*. University of California Press, Berkeley.

4. The translations from the *De rerum natura* are my own. I have used the second edition of the text given by Cyril Bailey (1922), Clarendon Press, Oxford.

5. H. J. Abrahams (1964), *Ambix* **12**: 44.

6. *De rerum natura* **lib. v**: 11. 793–6.

7. This is my translation of the French version quoted by Louis Pasteur in his lecture at the Sorbonne in April 1864. (1922–39), *Oeuvres de Pasteur* **2**: 328. Masson, Paris.

8. See note 2 Exercitatio 61.

9. Quoted by W. Clark (1992), 'The scientific revolution in the German nation', in *The scientific revolution in national context* (ed. R. Porter and M. Teich), p. 90. Cambridge University Press, Cambridge.

10. F. Redi (1668), *Experienze intorno alla generazione degl' insetti*. I have used the fifth impression published by the Stamperia di Piero Matini, Florence (1688).

CHAPTER 2

Flies and other insects

Francesco Redi (1626–98) (Fig. 2.1) was one of the most remarkable men in the history of biology. At a conference held in 1957 to mark the tercentenary of the foundation of the Accademia del Cimento,[1] Luigi Belloni, a distinguished Italian historian of science, called Redi 'the father of modern biology'. If that judgement is perhaps a little too selective, there is no doubt that Redi was one of its founding fathers. He was born in Arezzo,

Fig. 2.1 Francesco Redi (1626–98)

the son of Gregorio Redi, a renowned physician at the court of the Medici. After studies at Pisa and Florence, he entered into service at that court at the age of 21 and rose in due course to become head physician and superintendent of the ducal pharmacy and foundry. He remained at the Medici court to the end of his life.

Redi was a man of astonishing versatility—a celebrated poet, a profound classical scholar, a wayward (but impressive) etymologist, an excellent physician, and, above all, an experimentalist who changed the face of biology. The seventeenth century was not a high point in the history of Italian poetry but some critics consider him to be one of the foremost poets of that time. Eugenio Donadoni, in his popular *Brief history of Italian literature*, explains this rare conjunction of poetry and science by the fact that in seventeenth-century Florence, it was science rather than poetry that formed the centre of scholarly activity, so that if a poet did appear on the scene he was likely to be a scientist. Be that as it may, it is amusing to find that literary reference books make little or no mention of Redi's science; and scientific reference books, if they discuss Redi's literary work at all, usually limit their comments to the *Bacco in Toscana* (Bacchus in Tuscany). This is a dithyrambic poem in which Bacchus assesses the virtues and defects of the wines of Tuscany, with metric variations that are said to reflect increasing states of inebriation. He decides in the end that the best of them is the vintage that comes from Montepulciano:

> Bella Arianna con bianca mano
> Versa la manna di Montepulciano
> [Lovely Arianna with hand white as snow
> pours out the manna of Montepulciano]

I sonetti, a collection of 111 Petrarchan sonnets, again display Redi's technical virtuosity; here one also finds evidence of deeper emotional involvement, although, in these poems, it is very difficult to disentangle personal emotion from literary convention. One sees more of the man in the *Consulti* (medical consultations) and, of course, in his letters. The *Consulti* show Redi to be a learned, but eminently practical, physician whose advice is based on his own experience, not merely on doctrines handed down from antiquity. He is an advocate of the temperate life, strongly opposed to dietary excess. He seeks to help his patients by protecting them from charlatans and warning them of the perils of drug abuse. For him the practice of medicine is not only an intellectual exercise; there also are flashes of humanity that reveal a genuine concern for the welfare of his patients. An excellent example of this is a passage quoted by Belloni.[2] Finding that sugared solutions sometimes appeared to be active against worms, Redi comments: 'Would it not be a kindly remedy for poor thirsting boys afflicted with worms, to let them drink, out of handsome vessels, water simply sweetened with sugar or, better still, the sweetest, most fragrant lemon-water.'

Given the Florentine culture of the day it is not surprising to find that both Redi's prose and his verse are studded with classical allusions. In his case, this was not merely a superficial display of literary platitudes, but evidence of the most extraordinary classical learning. His references are often to obscure passages from obscure writers, which at that time can only have been the product of deep classical study and a singularly retentive memory. This, of course, was in the tradition of Florentine humanism, but Redi's knowledge apparently also extended, at least in some measure, to the languages of the Ancient East. He participated in the work of the Accademia della Crusca and was a major contributor to the dictionary that it produced. This academy was founded in 1583 with the aim of providing a critical appraisal of literary and linguistic work. The 'Crusca' (bolting sieve) was meant to symbolize the process of sorting wheat from chaff and was incorporated not only into the name, but also into the emblem of the academy. Like some latter-day linguistic academies, it eventually became a byword for pedantry. Redi offered a number of etymological proposals which turned out to be more imaginative than accurate, but which he continued to defend stoutly. He also wrote a history of the invention of spectacles which was very largely apocryphal, but which was accepted, at least in Italy, for some two centuries.

Whatever licence Redi allowed himself in his etymological and historical writing, this did not spill over into his scientific work. There we see an extremely sceptical attitude towards any assertions not based on meticulous observation; and his own painstaking experiments are a classic example of the new empiricism that set little store by received wisdom. During its brief life from 1657 to 1667, Redi was closely associated with the Accademia del Cimento (loosely translated as the Academy of Trial and Error), which, like its contemporary the Royal Society of London, refused to accept without evidence the authority of the written word. The academy's motto, to which Redi enthusiastically adhered, was *Provando e riprovando* (Test and test again). It was during the decade of the Accademia del Cimento that his major scientific work was done.

His *Observations on snakes*[3] appeared in 1664. In this, his first published scientific work, Redi shows, to begin with, that the snake's venom has nothing to do with the animal's bile, as was commonly believed. And then, in a masterly display of experimental skill, he demonstrates that the venom is harmless when taken by mouth and must be injected into the blood to exert its toxic effect. This demonstration rests on a complete acceptance of Harvey's discovery of the circulation of the blood, which was still being contested in print as late as 1655. In the knowledge that the blood does circulate, Redi recommends that in cases of snakebite a tight ligature be placed at a little distance from the bite in such a way that it impedes the flow of the envenomed blood to the heart. This proposal is one of the earliest examples of a therapeutic procedure based on physiological principles; it can reasonably be regarded as the origin of what we now call experimental toxicology.

Redi's account of his experiments on snake venom already shows the salient features of all his scientific writing. He presents an exhaustive review of the earlier literature, both ancient and modern, and does not hesitate to contradict the written word when his own experience disagrees with it. But this is done in a polite and dispassionate manner, without any trace of polemic. His own experiments are a classic illustration of *provando e riprovando*. The mode of delivery of the venom into the bloodstream is described in detail: the fangs in their retractable sheaths; the yellowish venom stored in the two glands that Redi found in all the snakes that he examined; the automatic injection of the venom when the snake bites. The investigation does not stop with the common European snakes that Redi has at his disposal; he also examines the venom of a deadly viper from the Dutch East

Indies. A bite from this reptile was said to be rapidly fatal, but its venom is still harmless by mouth. Nor does Redi limit his tests to one species of experimental animal; he feeds venomous snakes that he has killed to a dog, an owl, and another bird of prey, but none of them shows any ill effects. Finally, he examines excised tissues into which the venom has been injected; but these too are harmless when taken by mouth. And although they lie outside the scope of a treatise on snake venoms, he also reports similar observations on a deadly species of scorpion. Again, the venom has no toxicity when it is ingested. The extension of the work to scorpions makes it clear that Redi believes that his conclusion can be generalized: if, under natural conditions, a venom is injected directly into tissues, then it will be harmless if the tissues are taken by mouth.

Redi's masterpiece *Experiments on the generation of insects* appeared four years later; Italian and Latin versions were published contemporaneously. It is a masterpiece in two respects. First, it illustrates with a skill comparable to Harvey's the newfound primacy of experiment over the written word; and secondly, it provides a definitive answer to a question of the greatest biological importance. I have described how belief in spontaneous generation was gradually whittled away until it was generally accepted only in the case of insects. Insects, in the natural world then known, were the last bastion for the idea that animals could, under appropriate conditions, be generated from inanimate material without divine intervention. Redi's experiments reduced that bastion to rubble and, in doing so, left spontaneous generation without any empirical support.

The book has an epigraph in Arabic which Redi renders freely as a rhymed couplet:

> Chi fa Esperienze accresce il sapere
> Chi è credulo aumenta l'errore.
> [He who experiments increases knowledge
> He who merely speculates piles error upon error.]

Redi clearly nails the flag of empiricism to his mast. Then there follows a massively learned review of the ancient literature describing instances of spontaneous generation. This review includes all the examples that I have previously mentioned and many more. It is obvious that Redi views all these assertions with marked scepticism, but in his discussion of them he does no more than give a polite and dispassionate assessment of their plausibility.

We are not often given a frank first-hand account of what it was that prompted a scientist to embark on a completely novel series of experiments. And I think it is without parallel that the stimulus is said to be a well-known passage from a celebrated ancient author. Redi recounts how it was an episode in the Iliad that first raised his doubts about the doctrine of spontaneous generation. Here is the relevant passage from the beginning of Book XIX:

> Achilles, speaking to his mother, the Nereid Thetis, after the battle, says:
> 'but I am very much afraid that flies might in the meantime alight on the open wounds of the valiant son of Menoitios [Patroclus] and breed worms in them. Then his corpse would be defiled, for there is no life left in him, and his flesh will rot.'
> To which Thetis, the goddess of the silver feet, answers:
> 'Child, don't worry about that. I will see to it that he is protected from the cruel swarms of flies that prey on men slain in battle. For even if a whole year were to go by, his flesh would still be preserved, even better than it is now.'
> With these words she filled Achilles with courage again, and on Patroclus she shed ambrosia and poured red nectar down his nostrils so that his flesh might remain intact.

Redi says that on reading this passage he began to have doubts:

> What if it should turn out that all the grubs that you find in flesh are derived from the seed of flies and not from the rotting flesh itself? And my doubts were further strengthened by the fact that in all the generations that I bred, I always saw flies settling on the flesh before it became infested with grubs, and they were always flies of the same species as those to which the grubs eventually gave rise.

So, like Thetis, Redi decided to see whether, by warding off the flies, he could keep the flesh free from maggots.

The experiments that he designed to test this proposal were stark in their simplicity, but they were carried out with great care and with a sharp eye to possible objections. They were begun in the middle of July, when flies were abundant. Into four large, wide-mouthed flasks he put a dead snake, some river fish, four small eels from the Arno, and a slice of milk-fed veal. The mouths of the flasks were closed with paper tied down with string and were then well sealed. The same materials were placed in another four flasks but their mouths were left open. Within a short time, in the open flasks, the tissue samples were infested with grubs, and flies could be seen flying in and out of them; but in the sealed flasks the samples showed no sign of grubs, even though they were kept this way for several months and the contents had decayed, in some cases liquefied, and stank.

This experiment was repeated with different kinds of flask and for different lengths of time; and various other materials were put into the flasks, for example pieces of meat that had first been buried in the earth, or dead flies. Again, no grubs were produced in the sealed flasks, but, occasionally, a grub was seen on the outside trying to get in. Redi describes the grubs turning into eggs, but what he must have been observing here are later stages in the development of the grubs. However, he is not adamant about this. He considers it possible that some flies produce eggs while others produce live grubs; but he concedes that some authors say that they produce eggs only, which he neither affirms nor denies: 'io non affermo, e non nego'.

These experiments were decisive enough, but Redi was not quite satisfied. He foresaw an objection to them that was to be a bone of contention for more than two centuries. What if the generation of life required access to the open air? What if, in the sealed flasks, grubs were not formed because the air was cut off? Redi therefore prepared a second set of flasks in which the mouths were covered with a layer, sometimes two layers, of very fine muslin (*velo di Napoli*). He demonstrated that the air penetrated the muslin, but the cloth acted as an efficient fly-screen. Again, the screened flasks produced no grubs, but he noted that flies swarmed around the muslin, attracted by the smell. In some experiments, the flasks screened with a layer of muslin were placed in a box wrapped up in similar material. Redi noticed two grubs that easily got through the first layer of muslin and fell onto the second, which they partially penetrated. They gave rise to flies which buzzed within the box and laid eggs or grubs directly onto the muslin that sealed the flasks. These grubs generated more flies until the box swarmed with them. In all cases, Redi noted, the flies generated by the grubs were of the same species as those that had access to the decaying material. This was an observation of cardinal importance, for had the grubs been generated spontaneously, one would not have expected so stringent a relationship between the progeny of the grubs and the surrounding flies.

Redi then turned his attention to other insects, in the first instance bees. These, as previously mentioned, were traditionally thought to be generated from the decaying entrails of bulls. The entrails were therefore put into open and sealed flasks as before, and the same experiment was repeated. No grubs were produced in the sealed flasks, and the grubs that appeared in the open flasks generated flies, not bees. Similarly, it was shown that lice were not generated from decaying poultry, nor frogs from mud, nor scorpions from the earth, nor wasps from the flesh of horses, asses, or crocodiles, well-known instances of spontaneous generation cited in ancient literature. Eventually a very wide range of parasitic insects was examined (the treatise ends with detailed drawings of 29 different species, some of which are easily recognizable), and, in all cases, the results obtained with sealed and open flasks were the same. And this was true whatever the decaying material happened to be, animal or vegetable. Redi's historic conclusion was inescapable:

> Flesh and vegetable matter and anything else that is completely decayed or susceptible to decay play no part and have no other function in the generation of insects than to provide a site or a suitable niche for the grubs, or for their seeds, or for the eggs carried in by the animals or engendered by them at the time of parturition. And the grubs, as soon as they are born, find in this niche enough suitable food to nourish them.

Finally, with due caution, he makes the revolutionary generalization: he is 'inclined to think' that all living things are 'the faithful progeny of the plants and animals themselves and that these maintain the integrity of their species by means of their own seed'. This is perhaps the first comprehensive statement of genetic continuity in the history of biology.

Some authors consider that Redi defected from this general proposition in one case. In studying the formation of galls, especially crown galls in various kinds of oak, and similar excrescences in other plants, he is said to have reached the conclusion that the insects found within these growths were generated spontaneously. This is a misconception. Redi was not concerned here with spontaneous generation, but with whether the plant itself could generate an insect or whether this always entered the plant from without. This is a different question altogether. What is at stake here is not whether life can arise spontaneously from inanimate material, but whether vegetable matter can ever produce animal matter. This was a question of great philosophical importance at the time, for animals were self-evidently endowed with the spirit of sensitivity (*anima sensitiva*) whereas the spirit of plants was thought to be less elevated (*anima vegetale più ignobile*). To begin with, Redi was struck by the fact that the response of the plant tissue to attack by the insect was a proliferative one; infestation of a plant by an insect was, in general, destructive. Originally, Redi thought that the insect deposited on the surface of the plant some stimulatory fluid that induced proliferation of the plant tissue. His idea was that this stimulatory fluid was a 'fertile seminal liquor' ('fecondo liquore di seme') that had the power to penetrate into the deeper layers of the plant and thus induce a proliferative response. This was assumed to be necessary to provide adequate nourishment for the egg or grub.

About a decade after the first edition of Redi's *Generazione degl' insetti* appeared, Marcello Malpighi (1628–94) (Fig. 2.2), who, as described in more detail elsewhere,[4] was responsible more than anyone else for initiating the study of microanatomy, published the second part of his *Anatomes plantarum* (1679). This book contains a whole chapter devoted to galls ('De gallis'), and there is additional relevant information in the chapter entitled 'De plantis quae in aliis vegetant' ('On plants that grow within other plants'). Malpighi provided decisive evidence, amply illustrated, that what the insect deposited in the plant was not 'fertile seminal liquor'

Fig. 2.2 Marcello Malpighi (1628–94)

but an egg. The illustrations show with remarkable clarity the long retractable ovipositor of the gall-fly, which enables the insect to deposit its eggs in the deeper layers of the plant; larvae enclosed within swollen tissues in the thickness of the leaf; and, finally, the gall itself containing a larva at a further stage of development (Fig. 2.3). Malpighi's demonstration that it was the deposition of the gall-fly's eggs within the deeper tissues of the plant that gave rise to the galls was unanswerable; but why the tissue response should

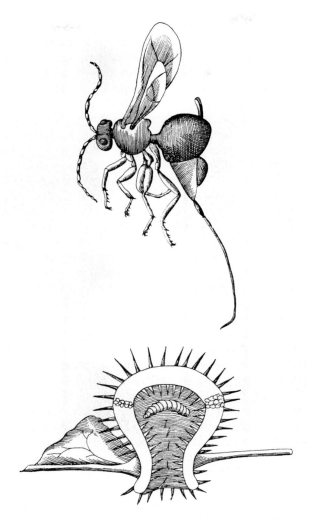

Fig. 2.3 Malpighi's drawings of the elongated ovipositor of the gall-fly (above) and of a gall containing a developing larva (below)

be proliferative remained unanswered. Malpighi's guess was better than Redi's, but it remained a guess. Malpighi proposed that the gall initially had the character of a morbid outgrowth produced by contact with the tip of the insect's ovipositor. This was thought to inoculate an 'extremely active and fermentative liquid' which induced 'fermentation and internal commotion' in the tender parts of the plant. The fermentation, Malpighi argued, provoked an inflow of nutritive juice which accumulated within the cells (*utriculi*), causing them to swell and eventually increase in number. The further development of the gall he envisaged as a reaction of the tissues of the plant to the growing larva, a form of symbiosis that adumbrated later ideas concerning the nature of parasitism.

Redi accepted without question Malpighi's demonstration of the role of the gall-fly and its larva in the generation of the gall. But he still had reservations about the explanation that Malpighi gave for the proliferative response of the plant tissues. Several observations remained unexplained. The galls always arose in growing parts of the plant, newly formed branches, or leaf junctions. They were not seen on the flat surface of the leaf. Redi never saw a single mature gall that did not contain a grub, but if he removed the gall in its earliest stages, he could detect no sign of a grub in it. Such galls, he claimed, never produced a grub or a fly. At later stages, he sometimes thought he saw an egg in the process of being formed, and sometimes a wizened grub that did not go on to maturation. Galls with more than one grub were also to be found. Some galls were formed in the spring, others in the autumn or the beginning of winter; some took two years to reach maturity. Galls of one species always enclosed the same kind of grub and produced the same kind of fly. These considerations led Redi to propose, not as an alternative to Malpighi's thesis but as an addendum to it, the idea that the plant itself might, at least in some cases, generate the insect.

Redi suggested, but no more than suggested, that the gall insect might be produced by the same natural life-force as that which enables the plant to produce its own fruit. What he actually said was:

and if I had to make my views public, it would be my belief that fruit, vegetables, trees and leaves become infested with grubs in two ways. One is that the grubs are delivered from without and, seeking food, gnaw their way into the pulp of the fruit or the pith of the wood. The other, which for my part I consider plausible, is that it would not be a major paradox to suppose that the life-force that produces the flowers and the fruit of living plants might be the same as that which produces the grubs within those plants.

It is clear, and was clear to him, that Redi was merely speculating, and it is a misconception to suppose either that he was supporting the doctrine of spontaneous generation or that he was opposing the views of Malpighi. Since it was not until the middle of the twentieth century that we began to have an inkling of what it was in the gall that initiated the proliferative response of the plant tissue, perplexity about this matter in the seventeenth century was, of course, inevitable.

Redi's last major scientific work was an exhaustive study of the parasites of animals and plants: *Observations on living animals that are found within the bodies of living animals*.[5] This work describes and illustrates an unparalleled range of parasites. Jules Guiart, a late nineteenth-century parasitologist, attempted to identify them and claimed to have done so in 108 cases, two-thirds of which were worms that colonize internal organs and one-third insects such as ticks and mites.[6] The list of animals Redi examined is, as one comes to expect of him, formidable: man, dog, hare, gazelle, hyena, fish, slugs, marine leeches, and 14 species of birds. Although plants are not mentioned in the title of the book, observations are included on some thirty kinds of flower. Being a doctor, Redi also tested a variety of remedies for the parasitic infestations of animals and man: onion, aloe, agaric, rhubarb, Peruvian bark, and orange blossoms. The scope of this work and the mentality that informs it fully justify the view of many Italian historians that the book is the foundation-stone of modern parasitology.

However, given Redi's abiding interest in the subject of spontaneous generation, it is at first glance surprising that there is little in the book about the origin of the parasites of internal organs such as worms or flukes. I think this is due to the fact that Redi, in his scientific work, made a clear distinction between what he had actually observed and what he thought was merely plausible. He did indeed envisage that these parasites engaged in some kind of complex life-cycle, but he could not determine what that cycle was; in one case, nonetheless, he came close. He thought that the hydatid cysts found in the liver of animals infested with tape-worms were embryonic forms of the worms. On the whole, how-ever, he did not dissociate himself from the view, generally held at the time, that intestinal worms were generated by the tissues of the host. Given his views on the plausibility of gall insects being generated in appropriate circumstances by the tissues of the plant, it is not inconsistent that he should have regarded it as even more plausible that animal tissues might similarly generate a new animal. As previously explained, a transformation of this sort is not, and was not thought by Redi to be, an instance of spontaneous genera-tion; but those who adhered to that doctrine continued to cite the obscure origin of intestinal worms as evidence in its support. This they did until the middle of the nineteenth century, when the com-plex life-cycles of these parasites began to be elucidated.

Redi did not for a moment concede that any of his work cast doubt on the divine origin of life or on any other Church doctrine. 'I remain', he writes, 'much more scandalized [scandalezzato] by certain authors who, on the strength of lies like this, reject the foundations and the exegesis of that highest of the Mysteries of the Christian faith, the resurrection of bodies at the end of the world.' Whether Redi was indeed so outraged, or whether he had his eye on the Inquisition when he inserted that largely irrelevant passage into his observations on insects, cannot now be deter-mined. It is, however, notable that throughout the centuries of controversy about spontaneous generation, in general, those with conventional religious beliefs most strongly opposed the idea. It was as if, whatever the evidence might be, believers wanted the origin of life to remain a divine mystery.

On 1 March 1698 Redi died, and his cadaver was taken back from Florence to his native city Arezzo. There his nephew, with the help of the Accademia della Crusca, erected an imposing tomb on which was sculpted:

> Francesco Redi Patritio aretino
> Gregorius Fratris filius.

On the title-page of his great book on the generation of insects Redi refers to himself as a 'Gentiluomo Aretino'. In the Latin translations of his works 'gentiluomo' becomes 'patritius' and sometimes even 'nobilis'. There is no doubt that he was a loyal and affectionate son of Arezzo and that he was conscious of his high birth.

Redi and Malpighi were fortunate in their disciples, particularly Bonomo and Cestoni in the case of Redi and Vallisnieri in the case of Malpighi. Giovan Cosimo Bonomo was a physician who was born in Leghorn in 1666 and died in Florence a mere thirty years later. Although active in Florence in the years before his premature death, he sometimes served as a physician in the galleys of the Grand Duke Cosimo III at Leghorn. Inspired by the work of Redi, he turned his attention to the human disease scabies. It had been known for some time that this skin condition was associated with the presence of mites, but it was generally thought that the mites were a consequence of the disease. No physician before Bonomo had proposed that the mites might be its cause. Working together with Giacinto Cestoni, an apothecary who was skilled in the use of the microscope, Bonomo was able to show that the mite reproduced by laying eggs ('minute, white and diaphanous') and that it had specialized mouth parts which permitted it to penetrate the skin. By means of these eggs, Bonomo succeeded in causing the mites to grow in dry cheese, thus demonstrating beyond doubt that they were not generated by the diseased skin, as was generally supposed.

These observations had immense medical importance, since they showed for the first time that a disease in man could be caused by the invasion from without of a parasitic organism. So surprising was this result thought to be that more than a century elapsed before it was incorporated into medical teaching. The transfer of the itch-mite from infected individuals to healthy ones was the first experimental evidence for the existence of a *conta-gium virvum*—a living agent capable of transmitting disease. This revelation eventually directed therapy away from internal remedies to medicaments that could be applied topically to rid the skin of the mites. The account of the experiments of Bonomo and Cestoni, *Observations concerning the itch-mites of the human body*,[7] was dedicated to Redi, who corrected and embellished the text before publication.

Antonio Vallisnieri (1661–1730) began his higher education at the Roman Catholic College in Reggio, but finding Aristotelian studies uncongenial, he moved to Bologna where he became a student of Malpighi who was then a professor there. Vallisnieri obtained his medical doctorate at Reggio in 1684 and returned to work with Malpighi for a further year. After a period of medical practice in Reggio he was appointed to a chair in Padua in 1706, the first chair in experimental medicine in that illustrious university. He remained in Padua for the rest of his life.

Vallisnieri attained a position of great eminence in the Europe of his day. He eventually became a Fellow of the Royal Society of London and was elevated to the hereditary nobility by Duke Rinaldo I of Modena. In the long view of history it is, however, difficult to avoid the impression that his fame rested as much on his advocacy as on his experimental originality. He completely accepted Malpighi's observations on the role of the gall-fly in the formation of the gall and generalized them to embrace all insects that parasitize plants, whether such plants produce a gall or not.[8] He was convinced that they attacked plants in order to lay their eggs. He argued, as Redi had done before him, that it was a law of nature 'that like always generates like', and that this genetic continuity was mediated via the egg. His views on the parasites of man and of domestic animals, especially intestinal worms, were less categorical. He did not believe that they were derived from earthworms or fruit worms, but, like all his contemporaries, he was unsure about what their origin was. He thought it possible that they were transmitted from mother to child, either in the fluids that nourished the foetus or in the mother's milk.

Much more given to sweeping generalizations of a semi-philosophical kind than Redi or Malpighi, Vallisnieri constructed a theory in which all life was said to be connected by a 'chain of being' so that man and animals were related. In addition to being an able advocate, Vallisnieri seems to have been an astute academic politician. In his inaugural lecture at Padua, given no doubt with a conservative audience in mind, he argued that recent studies did not subvert the tenets of ancient medicine, but, on the contrary, confirmed them. Later, however, he did not hesitate to criticize the assertions of traditional authorities. Indeed, the outlook that pervades his work is typical of seventeenth-century empiricism. He denied the existence of miracles and insisted that explanations of phenomena must rest on observation and experiment; occult interpretations were unacceptable. Everything in the natural world he regarded as a reflection of the mind of God.

Carlo Francesco Cogrossi, a physician from Crema, dedicated his work on cattle plague[9] to Vallisnieri. Cogrossi proposed that transmission of the disease was effected by a 'contagium vivum' made up of microscopic parasites. Vallisnieri, in his *Novel physical and medical observations*,[10] at once agreed and, in support of Cogrossi's proposal, added many observations of his own. This

Fig. 2.4 Antoni van Leeuwenhoek (1632–1723)

exchange between Cogrossi and Vallisnieri was an important early contribution to the theory of infectious diseases. Although the views that Vallisnieri espoused were clearly radical, he managed, apparently without difficulty, to placate the Inquisition. This talent is prominently displayed, for example, on the title-page of his *Novel physical and medical observations*, which was judged by the Venetian Inquisition to contain nothing against the Holy Faith. His collected works, including letters, were published posthumously by his son, also Antonio, in Venice in 1733.

While the empiricists in the universities of northern Italy— Florence, Bologna, Padua—were dismantling the defences of those who based their support of spontaneous generation on the obscure origin of insects, an extraordinary observation reported to the Royal Society in 1674 gave the theory a new and extended lease of life. Looking through lenses that he himself had ground, Antoni van Leeuwenhoek (1632–1723) (Fig. 2.4), a merchant in Delft, discovered a whole new world of living creatures incomparably smaller than insects. The mode of origin of these microscopic organisms was to remain a subject of heated controversy until the end of the nineteenth century. And there were even echoes of this controversy well into the twentieth.

Notes

1. L. Belloni (1958), *Francesco Redi biologo*. Domus Galilaeana, Pisa.
2. Ibid.
3. F. Redi (1664), *Osservazioni intorno alle vipere* (1678), *Observationes de viperis*. Leipzig.
4. H. Harris (1999), *The birth of the cell*. Yale University Press, London.
5. F. Redi (1684), *Osservazioni intorno agli animali viventi che si trovano negli animali viventi*. Florence.
6. J. Guiart (1898), 'Francesco Redi', *Arch. Parasitol.* **1**: 420.
7. G. C. Bonomo and G. Cestoni (1687), *Osservazioni intorno a' pellicelli del corpo umano*. Florence.
8. A. Vallisnieri (1713), *Esperienze ed osservazioni intorno alla origine, sviluppo e costumi dei vari insetti*. Padua.
9. C. F. Cogrossi (1714), *Nuova idea del male contagioso de' buoi*. Società Italiana di Microbiologia, Rome (1953).
10. A. Vallisnieri (1715), *Nuove osservazioni fisiche e mediche*. Venice.

CHAPTER 3

Microbes

What Leeuwenhoek initially saw were in all probability free-living protozoa. In a letter dated 7 September 1674 to Henry Oldenburg, secretary of the Royal Society, he reports having seen 'very little animalcules in fresh water'. His continuing observations on these and other microscopic organisms were communicated to the Royal Society in a series of letters. These have been translated into modern English by a group of Dutch scholars.[1] In his eighteenth letter, dated 9 October 1676, Leeuwenhoek describes many different sorts of creatures in rain-water including what are very likely to have been bacteria. 'I discovered very many exceeding small animalcules', he writes. 'Their bodies seemed, to my eye, twice as long as broad. Their motion was very slow, and oft-times round about.' Dobell, a scholarly biographer of Leeuwenhoek, considers that these organisms were probably bacilli.[2] Elsewhere Leeuwenhoek speaks of 'yet more oval creatures, and some exceedingly thin little tubes, which I had also seen many a time before'. It is, of course, impossible to be certain what, in modern terms, each of his descriptions corresponds to, but it is clear nonetheless that his observations revealed the existence of a whole panoply of microscopic organisms that ranged from bacteria to complex protozoa.

The widespread incredulity that greeted Leeuwenhoek's findings quickly gave way to a torrent of eager questions of which one of the most pressing was how these minute creatures were generated. That this should have been so is not surprising, for there were still many who adhered to the doctrine of spontaneous generation on traditional or philosophical grounds, but who could find no empirical support for their beliefs once the obscurity surrounding the generation of insects had been removed. The proposition that Leeuwenhoek's 'little animals' were generated spontaneously was asserted repeatedly and with vigour, so much so that an element of sharpness creeps into his own comments on the subject. Leeuwenhoek did not for one moment entertain the idea that the microscopic organisms he had discovered were produced from inanimate matter by spontaneous generation. His views are stated emphatically in several of his letters. Provoked by the claims of a certain unnamed gentleman, who maintained that vermin came from decaying matter and lice from sweat, Leeuwenhoek replies 'that this was pure imagination', and he 'is absolutely certain that it was just as impossible for a louse or a flea to come into being without procreation as it is for a horse or an ox, or some such animal to be born from the decay and corruption of a dung-heap'.[3] This, of course, is pure Redi, whose experiments Leeuwenhoek obviously accepted and, as shall be shown presently, probably copied.

We see the influence of Redi in other letters. Commenting on the observations of a 'certain prominent surgeon' who had removed a swelling from the leg of a lady and found the tissue full of small worms, Leeuwenhoek writes:

> On arriving home I examined the said Worms through the microscope, and I immediately decided that the Worms had come from Eggs which a Fly had laid on this mortifying part; and I also concluded that from these Worms, a species of Flies would again come forth, such as those that had laid the Eggs.[4]

And again:

> I have investigated the reproduction of the harmful little worm which gets into the corn, or grain, when they lie in the granary, and have sought how that little worm might be driven away, or removed from the corn. Our Bakers and corn dealers maintain that it is produced by the manure. But I state that it is just as impossible for a horse to be produced from a heap of manure, as for a creature gifted with mobility (which we call animal spirit) to emerge without procreation.[5]

In the following passage we see the argument generalized and developed into a coherent doctrine that was, and still is, the mainstay of those who believe in miracles but nonetheless reject the notion of spontaneous generation:

> However I definitely assert that, just as I have now clearly proved with regard to the Calenders; that they cannot originate except through propagation, so it must also be with all creatures that are endowed with movement (that which we call, in Animals, a living soul); for *however small it may be*, its first production depends upon the beginning of Creation; and if this were otherwise; namely if from the immobile substances such as Stone, Wood, Earth, Plants, Seeds etc, a body came forth that (as said above) was mobile, that would be a miracle and its production would once again be dependent upon the Great Almighty Creator.[6]

The two propositions contained in this categorical statement were, in the ultimate analysis, both acts of faith. The first was, of course, the orthodox belief that life was initially created by the miraculous intervention of a divine creator, and that only he had the power to re-create the miracle. The second proposition was more susceptible to experimental investigation, but it was not entailed by the observations that had so far been made. All the evidence discussed by Leeuwenhoek in the passages quoted deals with objects that are visible to the naked eye. In fact, what he has to say is no more than an elaboration and extension of the conclusions reached by Redi. But whether the 'little animals' that he had himself discovered were subject to the same rules remained problematical. Any opinion that he might have expressed about the origin of these creatures could only at that time have been an extrapolation from data that he did have to data that he did not. And, of course, he was in no position to draw any conclusions about the smallest creatures that he saw under the microscope, to say nothing of possible still smaller creatures that he could not see.

What Leeuwenhoek himself believed was that all the microscopic organisms that he had observed were generated by a process of normal reproduction analogous to copulation. He convinced himself that he could see the larger 'animalcules' coupling, distinguish male from female forms, and identify the offspring.[7] This interpretation of the appearances seen under the microscope was accepted by some investigators but not by others. Even after Abraham Trembley (1710–84) had shown that the multiplication of microscopic polyps was achieved by a process that split mature organisms into two, and Horace de Saussure (1740–99) and Lazzaro Spallanzani (1729–99) had demonstrated the probability that the copulation described by Leeuwenhoek was actually the movement of two daughter cells in the act of separating from each other, there were still eminent scholars who strongly supported Leeuwenhoek's interpretation. John Ellis (1710–76), the noted botanist, best known for his discovery that corals were made by a group of animals, continued to regard the motions that daughter cells exhibit when they separate as a 'copulatory movement'. He admitted that organisms could divide into two, but did not regard this as a normal mode of replication. He thought that the observed fission was actually fragmentation induced by shock. In support of his interpretation, Ellis emphasized that, in multiplying populations, 'animalcules' that did not divide outnumbered those that did by 50 to 1. Moreover, he claimed that he was able to discern young animalcules within older ones and still younger ones within the young ones. The idea that these microscopic creatures reproduced by some kind of sexual process was still current in the middle of the nineteenth century. Christian Gottfried Ehrenberg (1795–1876), in a book that became the standard work on the infusoria,[8] described in detail the male and female sexual organs that he thought he saw within the bodies of these creatures. Ehrenberg admitted that some of them divided into two but regarded this as an unusual form of reproduction. Indeed, one of the criteria that he used to classify protozoa was whether they divided into two or not.

But all these observations, assumptions, and arguments concerned the larger organisms in the complex microscopic population that Leeuwenhoek had revealed. Nothing was known, and very little was said, about the generation of the smallest creatures that could be seen. Leeuwenhoek was aware of this difficulty and tried to build a bridge between the small and the large organisms. He suggested that the former might aggregate to form the latter:

Observing that these creatures [the small ones] did augment into vast numbers, but not being able to see them increase in bigness, and neither having seen any such creatures floating in the water, I began to think whether they might not in a moment, as t'were, be composed or put together. But this speculation I leave to others.

Dobell[9] considers that this passage might indicate that Leeuwenhoek at one point entertained the idea of spontaneous generation, but given his repeated rejection of the idea elsewhere, this seems unlikely. The Dutch scholars who translated Leeuwenhoek's letters into English offer an alternative explanation. Since Leeuwenhoek sees the small organisms swimming near the surface of the fluid and the large ones near the bottom, and sees no intermediate forms, he suggests, but no more than suggests, that the small ones might 'in a moment' come together to form the large ones. Even though this suggestion is his own, Leeuwenhoek obviously doesn't think much of it, since he is prepared to 'leave it to others'. It seems to me that this interpretation of the text is much the more plausible and has, on the face of it, nothing to do with spontaneous generation in any case.

Leeuwenhoek was too committed an empiricist to suppose that the disagreement about the origin of his animalcules, especially the smaller ones, would be resolved by disputation alone. Although he was an observer rather than an experimentalist, he decided, no doubt under the influence of Redi, to put the matter to the test. The design of his experiment, like Redi's, was dramatically simple, but, unlike Redi, he failed to obtain the decisive result that he expected. The experiment and the results that it yielded were again reported in a letter to the Royal Society.[10] He filled two vessels with clean rain-water and pounded pepper, and then heated the mixture. The aperture of one of the vessels, which had been drawn out into a pointed end, was sealed by fire after the mixture had cooled. The other vessel was left open to the surrounding air. Three days later he took a little water from the open vessel and found it swarming with a variety of motile organisms. After a further 2 days he broke off the tip of the sealed vessel. The fluid, full of bubbles, was obviously under great pressure for it shot out of the aperture when the seal was broken. Leeuwenhoek confesses that he expected this vessel to be free of animalcules, but he again found a variety of them, including some that were so small that they were barely discernible.

An interesting retrospective diagnosis was made by M. W. Beijerinck (1851–1931), the eminent Dutch bacteriologist who was among the first to show that infections could be caused by entities so small that they could not be seen under the microscope and which passed through bacteriological filters. We now call them viruses. Beijerinck proposed that what Leeuwenhoek saw in his sealed vessels were bacteria that could grow in the absence of air or oxygen (anaerobic bacteria). While this might have been so, extended hindsight of this sort can only be a surmise. When he conducted his experiment, Leeuwenhoek could have had no inkling of what it meant for some creature to be 'anaerobic'; the role of oxygen in vital processes did not begin to be unravelled until more than a century later. If it was indeed the case that his experiment revealed the existence of anaerobic bacteria, then that discovery can only have been unwitting.

The importance of Leeuwenhoek's experiment was twofold. First, the experiment initiated the systematic attempts, which continued for two centuries, to settle by empirical enquiry whether his animalcules were generated spontaneously or not. And secondly, it laid the foundations for the methodology by which this enquiry was carried out. The idea of using open flasks and flasks shielded from the surrounding air was, of course, Redi's, but Leeuwenhoek adapted it for the specific purpose of investigating the origin of the microscopic organisms that he had discovered, and the long line of researchers who continued the investigation, whether they supported the idea of spontaneous generation or opposed it, all used flasks of one sort or another.

Some historians have set little store by the evidence provided by these experiments with flasks.[11] There appear to be two reasons for this. The first derives from a mistaken application of the principle in logic that no empirical observation or set of empirical observations can entail a universal negative. The argument here seems to be that one cannot conclude that spontaneous generation does not exist from any number of demonstrations that it has not occurred in a particular set of circumstances. The second reason is that none of the experiments conducted with flasks was flawless, and the flaws are seen not only now, with historical hindsight, but were also seen, and were eagerly seized upon, by contemporaries. The literature on spontaneous generation does not contain a single report in which it is claimed that an extensive series of experiments yielded an unblemished set of results. There is always the occasional flask that supports the growth of micro-organisms under experimental conditions that decisively preclude such growth in the vast majority of similarly treated vessels. If one accepts the view that a hypothesis can be falsified by a single counter-example, then these exceptional flasks may be thought to invalidate the conclusion that, under experimental conditions, spontaneous generation does not occur.

Both arguments require further discussion. The great majority of the scholars who turned their attention to the question of spontaneous generation had received a thorough humanistic education and were as well aware as we are of the impossibility, in logic, of proving a universal negative. But this was not at all what they were trying to do. They sought to devise experimental conditions that eliminated, or at least minimized, extraneous sources of error. More often than not, an investigation was undertaken to examine a claim made by someone else that spontaneous generation could indeed be observed under certain specified conditions; and as the investigation proceeded, these conditions were refined, new variables delineated, and new sources of error revealed. All this is good orthodox scientific practice and has nothing to do with proving a universal negative. Indeed, it was not until the second half of the nineteenth century, when in a public lecture Pasteur asserted that, in his view, spontaneous generation did not occur under *any* experimental conditions, that a conclusion even resembling a universal negative is to be found in the literature. Pasteur's predecessors, in discussing their negative findings, were always circumspect and rarely extrapolated beyond the confines of their own experiments.

The proposition that a single counter-example falsifies a hypothesis has little relation to experimental reality. To begin with, it may be the counter-example that is false, not the hypothesis. The counter-example may be a simple experimental error, or it may, with a little modification, be made compatible with the hypothesis. What actually happens is that the experimenter assesses the weight of the counter-example. He makes a judgement about whether it can simply be dismissed on technical grounds, whether it does indeed seriously undermine the hypothesis, or whether, while not actually undermining the hypothesis, it does nonetheless raise difficulties that require further exploration. In the case of the historical experiments in which flasks of various kinds were used to examine the question of spontaneous generation, it is a remarkable fact that it was the last of these judgements that was almost invariably made. Rarely was an aberrant result dismissed out of hand. The unintentional end result of this habitual caution was a major advance in our understanding of how microorganisms manage to withstand even the most inclement of environments.

Clearly, the history of the controversy surrounding the idea of spontaneous generation cannot be written without due consideration being given to the background against which this controversy took place. Any number of theologians, philosophers, and scientists held strong views on the subject; and these views were, of course, guided by religion, politics, and a host of other influences. But one must distinguish between an expression of opinion that is not based on any empirical evidence and a conclusion drawn from experiment. Many eminent men took sides not because they were convinced by the experimental data, one way or the other, but because belief or disbelief in spontaneous generation was more in accord with their own world-view. Nonetheless, it was not this ever-present chorus of unsupported opinion that eventually settled the debate. That was resolved by a long line of meticulous experiments; and if one chooses to ignore these experiments or belittle them, then one has chosen to write about the singer, but not the song.

Notes

1. *The collected letters of Antoni van Leeuwenhoek.* Swets and Zeitlinger, Amsterdam (1948). Further information about Leeuwenhoek and his letters is given in Harris, *The birth of the cell.* See Chapter 2, note 4.
2. C. Dobell (1932), *Anthony van Leeuwenhoek and his little animals.* Bale and Danielsson, London.
3. Letter 14 May 1686. See note 1.
4. Letter 17 October 1687. See note 1.

5. Letter 18 September 1691. See note 1.
6. Letter 6 August 1687. See note 1.
7. References to the letters containing Leeuwenhoek's views on the generation of his animalcules are given in Harris, *The birth of the cell*, pp. 54–6. See Chapter 2, note 4.
8. C. G. Ehrenberg (1838), *Die Infusionsthierchen als vollkommene Organismen*. Voss, Leipzig.
9. See note 2.
10. Letter 14 June 1680. See note 1.
11. See, for example, J. Farley (1974), *The spontaneous generation controversy from Descartes to Oparin*. Johns Hopkins University Press, Baltimore.

CHAPTER 4

The battle of
the flasks begins

Leeuwenhoek did not believe in spontaneous generation, but the vehemence of his statements about it indicates that there were many who did. And their position was greatly strengthened by the fact that his own attempt to dismiss the idea by an experiment similar to Redi's had failed. He had expected that there would be no growth of his animalcules in the flask shielded from the surrounding air, but there was. Leeuwenhoek does not report having tried the experiment again or varied it. So, as matters then stood, there was nothing in his observations that made belief in the spontaneous generation of microbes more difficult. People believed what they wanted to believe, unconstrained by any experimental evidence. But what they believed was not determined simply by their faith. Both John Turberville Needham (1713–81), who remained a passionate advocate of spontaneous generation for decades and who appears to have been the first to provide experimental evidence in its support, was a Catholic priest; but so was Lazzaro Spallanzani (1729–99), who vigorously opposed it.

Needham was the son of recusant parents who had fled to France to escape the persecution of Catholics in England. He was educated at the college established at Douai for the children of such exiles and later taught rhetoric there. He was ordained a secular priest in 1738. His scientific interests appear to have been motivated largely by his desire to defend the faith or at least his interpretation of it. He was widely known in educated circles for the dispute that he had with Voltaire about miracles. He was made a Fellow of the Royal Society in 1747 and of the Society of Antiquaries in 1761. In 1768 he founded a society in Brussels, the Imperial Academy, that eventually became the Académie royale des sciences, and in 1773 he became the first director of the Academy. His was a completely expatriate bilingual life and it was in Brussels that he died.

The work on spontaneous generation for which Needham is now remembered appears to have been initiated in 1748 by an invitation from Buffon in Paris. Georges Louis Leclerc Buffon (1707–88) was one of the great figures of French science in the eighteenth century. In 1739 he was appointed director of the Jardin du Roi and the Royal Museum, and there began assembling the material for his monumental *Histoire naturelle*, which was published between 1749 and 1767 in 44 volumes. The aim of this massive work was to present and discuss all that was then known in the field of natural history. Buffon was eventually ennobled by Louis XV. His views on spontaneous generation were complex and sophisticated: he had no doubt that it occurred, and envisaged that, after death, the decomposed tissues of living creatures released organic 'molecules' which were endowed with the ability to reassemble and again form living creatures. This reassembly was thought to be guided by templates (*moules*) which were characteristic of the creature from which they were derived; but organic molecules emanating from one creature could reassemble on the template of another thus producing an organism of a different kind. Buffon proposed that the overall process was governed by chance.

In inviting Needham to Paris, Buffon hoped to be able to obtain experimental evidence in support of his theories. It appears to have been Needham who carried out the experiments. Buffon accepted the results and incorporated them into his subsequent writings. In his own account of the work, Needham sought to establish a clear distinction between his contributions and those of Buffon. In part, this was probably due to the almost universal desire of collaborators to ensure that their work is not buried in the writings of someone else. In his correspondence, Needham refers to '*my* system of spontaneous generation and epigenesis'. But his views on spontaneous generation did differ in some important respects from those of Buffon. In particular, he did not accept the idea that the process was governed by chance. The two most important sources of information about Needham's experiments are the paper he communicated to the Royal

Society in 1748[1] and his comments on the work of Spallanzani. Spallanzani's account of his own now classic experiments on spontaneous generation was essentially a polite, but devastating, critique of the theories of Needham and Buffon. It appeared in 1765 and bore the title *An account of microscopical observations concerning the theory of generation of Messrs Needham and Buffon.*[2] Needham translated this into French,[3] but added notes criticizing Spallanzani's experiments and defending his own position. These notes throw light on the views that Needham still held some two decades after the work described in his communication to the Royal Society of 1748.

The 1748 paper begins with a review of the opinions then current on the subject of spontaneous generation. Needham concedes that 'the opinion of pre-existent Germs has prevailed', but rejects this view because he regards it as based not on evidence but on analogy. In particular, he has in mind the experiments of Redi and Malpighi on insects, and he attacks the notion that 'microscopical animalcules were generated from Eggs transported through the air, or deposited by a *Parent Fly*, invisible to the naked Eye, or even assisted with microscopes'. Leeuwenhoek, he admits, had certainly carried out 'very nice Operations upon extremely minute subjects', but his conclusions concerning the mode of generation of these microscopical creatures was, in Needham's view, an example of the errors inherent in conclusions drawn from analogy. Since he intends to argue in favour of spontaneous generation, Needham naturally also rejects 'various experiments that seem'd to prove every Animal, every Plant, descended from Individuals of the same species'. In short, he proposes to 'take as little Notice as may be, in this short Summary, of the almost inevitable Mistakes others may have made in this Matter before me'.

Needham's initial experiments appear to have been instigated by, and were carried out in collaboration with, Buffon. 'Thus did our Enquiry commence upon Seed Infusions from a Desire M. de Buffon had to find out the organical Parts.' An infusion of almond seeds was stirred and enclosed in 'Phials with Corks'. Eight days later 'a languid motion in some of the seed-particles' was noticed. Needham observes that there are 'Compound Bodies in Nature, not rising above the Condition of Machines, which might yet seem to be alive, and spontaneous in their Motions'. However, in the almond seed infusion a 'distinct Atom would often detach itself from others of the same or less dimensions', but the motion of such 'Atoms' was not judged to be 'spontaneous'. A few days later, however, the outcome of the experiment was clear. 'The result of our first Observation was that tho' the Phials were close stopped, and all communication with the exterior Air prevented, yet in about fifteen Days Time, the Infusions swarm'd with Clouds of moving Atoms, so small and so prodigiously active.' Buffon no doubt thought that these motile Atoms were the organic 'molecules' that could reassemble into more complex living creatures, but at this point there appears to have been a disagreement between the two men over the interpretation of the results, for thereafter Needham describes only his own experiments and gives his account of them in the first person:

> For my purpose therefore I took a Quantity of Mutton-Gravy hot from the Fire and shut it up in a Phial, closed with a Cork so well masticated [sealed with mastic], that my Precautions amounted to as much as if I had sealed my Phial hermetically... I neglected no Precaution, even as far as to heat violently in hot Ashes the Body of the Phial.

But after some days, 'My Phial swarmed with Life, and microscopical Animals of most Dimensions, from some of the largest, to some of the least.'

Needham repeated the experiment with three or four scores of different infusions, in all of which he observed 'the same Phaenomina with little Variation'. He also studied the constituents of male semen, where he saw 'moving globules, and trailing behind them something like long Tails... These *vegetative Powers*', he concludes, 'from the very Beginning of my Observations I had found to reside in all Substances animal or vegetable, and in every part of those Substances'.

The events that occurred in an infusion of pounded wheat are described in more detail:

> If any particle was originally very small and spherical, as many among those of the pounded seeds were, it was highly agreeable

to observe its little star-like Form with Rays diverging on all Sides, and every Ray moving with extreme Vivacity. The Extremities, likewise of this gelatinous Substance, exhibited the same Appearances, active beyond Expression, bringing forth, and pairing continually with moving progressive Particles of various Forms, spherical, oval, oblong and cylindrical, which advanced in all Directions spontaneously, and were the true microscopical Animals so often observed by Naturalists... In the Infusion of pounded wheat, the first appearances, after an exhalation of volatile Parts, *as in every other Infusion*, were the second or third Day Clouds of moving *Atoms*, which I suppose to have been produced by a prompt Vegetation of the smallest and almost insensible Parts, and which required not so long a Time to digest as the more gross.

These 'Atoms' settled and remained 'absolutely inactive till about fourteen or fifteen Days' when they united to form filaments —'Zoophytes all, and swelling from a Force lodged within each Fibre'. These filaments at first had a 'vermicular Motion' but eventually reached 'a higher Degree of Maturity and Perfection'.

Needham was aware of Abraham Trembley's observations on microscopic polyps.[4] Trembley had decisively shown that multiplication of these polyps was driven by a process that involved the mature polyp dividing into two. Needham is prepared to concede that there might be something special about the generation of polyps, but believes nonetheless that they were originally formed by spontaneous generation as he had previously described it.

Needham's summary ends with a series of conclusions designed to make his 'System of Generation' clear:

It seems plain therefore, that there is a vegetative Force in every microscopical Point of Matter, and every visible Filament of which the whole animal or vegetable Texture consists. And probably this Force extends much farther... This is not only true of all the common microscopical Animalcules, but of the spermatic also; which, after losing their Motion, and sinking to the Bottom, again resolved into Filaments, and again gave lesser Animals. Hence it is probable, that every animal or vegetable substance advances as fast as it can in its Resolution to return by a slow Descent to one common Principle, the source of all, a kind of universal *Semen*; whence its Atoms may return again, and ascend to a new Life...Nor indeed can there be a stronger Argument derived from any System of Generation whatsoever, of an All-wife Being, All-powerful, and All-good, who gave to Nature its original Force, and now presides over it, than from the Consideration of an exuberating ductile Matter, activated with a vegetative Force...

Needham's experimental work thus has two components. The first is his repeated demonstration, 'with little Variation', that infusions of organic matter heated and then shielded from the surrounding air nonetheless generate living forms. And the second is his microscopical examination of the process by which this generation might occur. It is the first of these that has the greater historical importance. In the eighteenth century observations made with the microscope were regarded with some scepticism; and Needham's conclusion that the replacement of one population of microscopic objects by another meant that the one had been transformed into the other, had its opponents even then. But those who believed or wanted to believe in spontaneous generation eagerly welcomed Needham's experiments with sealed phials, especially as Leeuwenhoek, who did not believe in it, had obtained the same result—the growth of microorganisms in an infusion that had been heated and shielded from the surrounding air. On the other hand, those who did not believe, or did not want to believe, in spontaneous generation continued to regard this result as an aberration due to some as yet unidentified error in technique. This polarization of opinion persisted to the end of the nineteenth century, and one can find echoes of it in the twentieth.

Lazzaro Spallanzani (1729–99) (Fig. 4.1) was one of the most skilful and most versatile experimentalists of his day. He made major contributions to many areas of biology, but is now perhaps best known for his meticulous dismantling of the evidence that had so far been adduced in support of the doctrine of spontaneous generation. He was born in Scandiano in the duchy of Modena and attended a Jesuit school in Reggio Emilia. After law studies at Bologna, he took minor orders and was ordained

Fig. 4.1 Lazzaro Spallanzani (1729–99)

a priest. Although attached to two congregations in Modena, he appears to have performed his priestly offices rather irregularly, but did officiate at Mass from time to time, even in later life. His initial academic appointments were in Reggio, where in 1757 he was elected to the chair of mathematics and physics. From 1763 to 1769 he held the chair of philosophy at Modena and then moved to the chair of natural history at Pavia where he remained to the end of his life.

Spallanzani's analysis of spontaneous generation came in two parts. The first was the previously mentioned *Saggio* of 1765,[5] which was a frontal assault on the theories of Needham and Buffon; the second was his answer to the notes that Needham had added to his translation of the *Saggio* into French. Both parts are combined in the *Observations and experiments concerning the animalcules of infusions*,[6] the final version of which is to be found in the *Treatises on animal and vegetable physiology*, published in Modena in 1776[7] and translated into French by Jean Senebier in 1777.[8]

Spallanzani has two principal objections to Needham's experiments with sealed phials. The first is that Needham did not heat the vessels long enough to destroy the seeds of the animalcules that eventually appeared in the infusion. The second is that the phials were sealed only with corks, which Spallanzani regards as porous. It is possible, but unlikely, that the corks used in Italy in the eighteenth century were more permeable than they are today, but, be that as it may, Spallanzani overlooks the fact that Needham sealed his corks thoroughly with mastic. In any case, Spallanzani decides to repeat Needham's mutton gravy experiment, but with two modifications. He seals the vessels hermetically by flaming the neck until the glass fuses, and he immerses them in boiling water for an hour. In all, he treats 19 vessels in this way and in none of them does he find any trace of animalcules when the vessels are eventually opened.

This result might well be thought to have put an end to Needham's theory, but this was not to be the case. Needham's reply is to be found in the notes that he added to his translation of the *Saggio* into French.[9] Needham claims, writes Spallanzani, that I 'greatly reduced, perhaps destroyed, the vegetative force of the infusions by exposing my vessels to the action of boiling water for an hour'; and also 'that I greatly damaged the elasticity of the air that remained enclosed in the vessel by subjecting it to the exhalations of the water and the intensity of the flame'. It is hard to know whether Needham genuinely believed in these objections or whether he was simply casting about for arguments that might save his tottering case. Needham's objections seemed to Spallanzani, and seem to us now, to be contrived, but they initiated a long critical tradition in which any experiment that involved the use of shielded flasks was rejected on the ground that the manipulations had in some way damaged the enclosed air.

Spallanzani explored Needham's objections with great care. He used infusions made from a wide range of different organic materials: white beans, vetch, buckwheat, barley, maize, mallow seeds, white beet, and egg yolk because this was known to harbour a large number of animalcules. To avoid variations in boiling point due to fluctuations in atmospheric pressure, all flasks were boiled at the same time, for periods ranging from half an hour to two hours. The vessels were numbered to avoid confusion and left open to the surrounding air. For microscopy several samples were taken from each infusion. If the infusion was too thick for microscopy, it was diluted with distilled water, as ordinary water often teemed with animalcules. Thirty-two infusions were tested. The results were variable: maize infusions generated animalcules that were smaller in size and number the longer they were boiled; but infusions made with beans, vetch, barley, and mallow seeds contained animalcules that appeared healthier after having been heated for two hours. Different sorts of animalcules could be found in the one infusion and, in some cases, one sort was present at the beginning of the experiment and another at the end. Spallanzani's conclusions to this set of experiments is as follows:

> The clear outcome of these experiments is that prolonged boiling of the infusions made with seeds does not stop the animalcules hatching, and although the experiment with boiled maize does not support this conclusion, the results given by four other infusions are in complete agreement with it, as can be seen.

But Spallanzani does not stop there. He tests additional infusions, peas, lentils, beans, and hemp, and finds that, in these too, the animalcules are most abundant in the flasks that were boiled longer. Why this should be so he does not know, but suggests that longer boiling might cause more effective disintegration of the seeds.

While these experiments clearly demonstrated that, in flasks open to the air, the growth of animalcules was not inhibited by prior exposure of the infusions to the temperature of boiling water, Spallanzani was curious to know what effects higher temperatures might have. He roasted 11 kinds of seed in a coffee-bean roaster and tested infusions made from them. Again there was growth of animalcules in the open flasks. And this was still true when the roasting temperature was 110 °C, as measured with a Réaumur thermometer, or when the seeds were burnt on a brazier, and even when they were reduced to charcoal in the flame of a reverbatory furnace. Spallanzani concludes that Needham's 'vegetative force', which was supposed to endow the basic components of living matter with the power to reassemble and reform more complex creatures, was pure imagination.

Spallanzani then turns to Needham's second objection, namely that the intense heat of the flame coupled with the 'exhalations of the water' had destroyed the elasticity of the air. This objection he meets in the following way. By heating the infusion before sealing the flask, he shows that when the seal is broken, the outside air comes whistling in; and the direction of the airflow he confirms with a lighted candle. He then draws out the neck of the flask until only a capillary opening remains. This he seals quickly so that the air inside the flask can be presumed to be at the same density as the air outside. When he now breaks the seal, there is no whistling. If the sealed capillary is broken after a period of 11 days, during which some growth of animalcules has occurred, the air in the flask rushes out, the direction of the airflow again being confirmed by a lighted candle. It was thus clear that the air in the flask had not lost its elasticity.

Both of Needham's objections were thus irretrievably disposed of, but Spallanzani was still not satisfied. The unlikely possibility that air might have passed through the heated glass was examined by repeating the crucial experiments in metal containers themselves sealed with metal, but the results were no different. Spallanzani noticed, however, that the larger animalcules in his infusions were much more susceptible to heat than the smaller. Detailed timing experiments showed that a very brief exposure to heat killed off the largest, but some of the smallest survived prolonged heating. This observation cast further doubt on the procedure used by Needham for killing off pre-existing microorganisms, but, more important, it revealed that some of them were remarkably heat-resistant. Spallanzani concludes that it is the *seeds* of these organisms that are heat-resistant and discusses the possibility that there might be a range of lethal temperatures for different seeds. In support of this view, he quotes Henri Louis

Duhamel de Monceau (1700–82), an eminent French biologist, who in 1765 showed that wheat could be germinated after being heated in a stove to a temperature $10\,^\circ$C higher than that of boiling water.[10]

The resistance of microorganisms to inclement conditions continued to intrigue Spallanzani. He investigated the effects of very low temperatures as well as high temperatures, and found that some animalcules could continue to multiply even at the temperature of ice. The effects of electricity, vacuum, and a variety of chemical agents were also examined. Although bacterial spores were not discovered until the second half of the nineteenth century, it is abundantly clear that Spallanzani was well aware that some animalcules or their 'seeds' were resistant to the most extreme conditions.

To modern eyes Spallanzani's investigation is a model of precision and completely convincing, but in the eyes of his contemporaries, and for a century afterwards, his experiments were regarded as inconclusive. One reason for this was, of course, the persistence of preconceived opinions, usually based on some philosophical or religious view of how life originated in the first place. But perhaps more important was the fact that it was much easier to achieve Needham's result than Spallanzani's. Experiments little different from Needham's continued to be conducted until the end of the nineteenth century and continued to reveal the growth of microorganisms in sealed vessels. Spallanzani's main criticism of Needham's experiments was that unless stringent precautions were taken to eliminate pre-existing microorganisms and contamination of the vessel by microorganisms in the surrounding air, it was very likely that growth would take place, but growth under such lax conditions was an experimental artefact and no evidence at all of spontaneous generation. Few, if any, of Spallanzani's contemporaries were prepared to go to the lengths that he had reached to eliminate pre-existing organisms and exclude contamination from without. So, in the minds of many, Needham's conclusions and those of Spallanzani were both plausible interpretations of the observations that had been made, and opinion remained divided.

A further complication was introduced by Spallanzani's observations that temperatures substantially higher than that of boiling water were needed to kill some organisms. Although this observation initiated a line of work that eventually proved to be of great importance to modern microbiology, what temperature was actually required to kill all pre-existing organisms remained uncertain. It was not appreciated at the time, nor does it seem to be even today,[11] that this uncertainty attaches to Needham's experiments but not to Spallanzani's. Needham's results could, as Spallanzani had suggested, have been due to the survival of organisms not killed by the temperature used, but what the lethal temperature for these organisms might have been is not relevant to the experiments that led Spallanzani to his conclusions about spontaneous generation; for, in that series of experiments, there were no surviving organisms. To cast doubt on, or dismiss, all experiments conducted with sealed flasks because of uncertainty about lethal temperatures reveals an incomplete understanding of what these experiments were intended to show and what they were capable of showing.

Finally, Spallanzani's experiments were met by the almost unanswerable objection that there was some, unspecified, property of the air that was indeed destroyed by heating, even if the notion that it was the air's elasticity had been eliminated. It was the quality of the purified and shielded air that, for a century after Spallanzani's death, preoccupied both the supporters and the opponents of spontaneous generation.

Those who, like Leeuwenhoek and Spallanzani, rejected spontaneous generation were nonetheless under an obligation to provide some alternative explanation of how microorganisms multiplied. As previously discussed, Leeuwenhoek believed that multiplication took place by means of sexual conjugation basically similar to that seen in animals and man. Spallanzani rejected this and, under the influence of Abraham Trembley's work on microscopic polyps, proposed that microorganisms multiplied by binary fission, that is by the division of the mature cell into two. It was Spallanzani's view that eventually prevailed, but, in 1946, the discovery was made that bacteria, the smallest of Leeuwenhoek's animalcules, could, under certain circumstances, engage in sexual conjugation also, even if the process bore little resemblance to sexuality in animals and man.[12]

Notes

1. J. Needham (1748), *Phil. Trans. Roy. Soc.* **45**: 615.
2. L. Spallanzani (1765), *Saggio di osservazioni microscopiche concernenti il sistema della generazione de' Signori di Needham e Buffon.* Modena.
3. J. Needham (1769), *Nouvelles recherches sur les découvertes microscopiques et la génération des corps organisés.* London and Paris.
4. A. Trembley (1744), *Phil. Trans. Roy. Soc.* **43**: 169.
5. See note 2.
6. *Osservazioni e sperienze intorno agli animalculi delle infusioni.* See note 7.
7. L. Spallanzani (1776), *Opusculi di fisica animale e vegetabile.* Modena.
8. J. Senebier (1920), *Observations et expériences faites sur les animalcules des infusions par Lazare Spallanzani.* Gauthier-Villars, Paris.
9. See note 3.
10. H. L. Duhamel de Monceau (1765), *Supplément au traité de la conservation des grains.* Paris.
11. See Chapter 3, note 11.
12. J. Lederberg and E. L. Tatum (1946), *Cold Spring Harbor Symp. Quant. Biol.* **11**: 113.

CHAPTER 5

Materialism, for and against

At no time in the two hundred years that followed Leeuwenhoek's discovery of his 'little animals' was there a lull in the controversy surrounding spontaneous generation. It was a subject that engaged the minds of theologians, philosophers, and natural scientists of every kind. The literature is studded with some of the most illustrious names in the intellectual life of Europe. There were, as I have explained, two fundamental questions, distinct but often confused: how, in the first place, did life on earth begin; and, given the appropriate conditions, could living organisms be created from inanimate matter at will. For those who accepted a literal interpretation of Genesis or of any of the creation myths of other ancient religions, the answer to the first question was readily to hand: life was the product of divine intervention. But even for true believers, it remained uncertain whether that intervention occurred only once, or whether it was repeated from time to time. An omnipotent God could, of course, create life from inanimate matter whenever he chose; but did he so choose? Those who regarded creation myths as metaphors were in greater difficulty, especially those who sought to eliminate divine intervention altogether.

The French materialists of the eighteenth century, who believed, after Descartes, that nothing existed but matter and its movements, were all obliged to postulate some form of spontaneous transition from the inanimate to the animate. Descartes himself believed that all that was necessary to produce living organisms from putrefying organic material was that it should be heated and shaken.[1] Etienne Bonnot de Condillac (1715–80), an ordained priest known in his day as l'abbé de Condillac, carried physicalism a stage further. Under the influence of John Locke, he argued that all human mental faculties such as memory and imagination could ultimately be traced to physical sensations, and that, by the exercise of precise logical reasoning, it was possible to move from the principles that govern the material world to those that govern society, and even to encompass the idea of God and the immortality of the soul. Georges Cabanis (1757–1808), a physician and philosopher in the same materialist tradition, regarded the spiritual side of human nature not as a separate entity, but as a function of physical nature. Since his philosophical position excluded explanations that were not grounded in material reality, it comes as no surprise that, in explaining the origin of life, he should have advocated spontaneous generation.

A typical eighteenth-century transition from belief to disbelief is seen in the development of Denis Diderot (1713–84), the driving force behind l'Encyclopédie, the vast encyclopaedic dictionary in which the accumulated knowledge of the eighteenth century was assembled. Diderot was educated by Jesuits but decided at an early stage not to enter one of the regular professions or the Church, rather to devote himself to literature. It was not long before his literary activities drew him away from religious orthodoxy, and by the time, in his thirty-third year, he published his Pensées philosophiques (1746), condemned in due course by the parlement of Paris, he had drifted into deism and viewed with distaste explanations of natural phenomena that involved supernatural intervention. The deism soon hardened into radical materialism, and history now sees Diderot as the archetypal eighteenth-century materialist. The most explicit account of his mature views on spontaneous generation is to be found in his Rêve de d'Alembert (d'Alembert's dream),[2] which was not intended for publication and was, in the event, published posthumously. Jean le Rond d'Alembert (1717–83) was a distinguished mathematician and physicist whose philosophical position was even more radical than that of Diderot. An unrelenting opponent of religion and the priesthood, he was for some years Diderot's principal collaborator in the preparation of the Encyclopédie. The Rêve is an uncompromising exposition of the materialist view of man. Like Descartes, Diderot proposed that life had its origin in an inert organic fluid that was converted into a living organism by the action of heat. Most contributors to the *Encyclo-pédie* (the *Encyclopédistes*), having rejected supernatural inter-vention, adopted, as a plausible alternative, some variant of spontaneous generation.

Eighteenth-century materialists saw spontaneous generation as a recurrent and altogether natural process. They were not there-fore faced with one problem that confronted those who believed that the creation of life required divine intervention. Materialists did not have to decide whether the creation of life had occurred only once or whether it occurred repeatedly. This question did not become a subject of serious scientific debate until doubts began to emerge about the fixed nature of species. In the eight-eenth century both believers and disbelievers generally assumed that a particular species, animal or plant, no matter how or when it was created, underwent only minor variations thereafter. Towards the beginning of the nineteenth century, however, fragments of evidence began to appear that led some naturalists to question whether this was necessarily so.

Nowhere are the issues at stake more clearly exposed than in the protracted, and often acrimonious, controversy between Cuvier and Lamarck. Jean-Baptiste Lamarck was born at Bazentin in Picardy in 1744, the last of 11 children in a family of aris-tocratic lineage, but modest means. He was educated in a Jesuit college at Amiens, and, after the expulsion of the Jesuits from France in 1761, served for some years in the army. An injury forced him to resign and was, in some measure, responsible for his eventually taking up a scientific career. He had at least three wives, begat eight children, and ended his life in great poverty and totally blind (Fig. 5.1). His early work, begun while he was still a soldier, dealt mainly with the identification and classification of plants. It attracted the attention of Buffon through whose influ-ence Lamarck was elected into the botanical section of the Aca-démie des sciences and was eventually given a salaried position at the Jardin du Roi.

In the post-revolutionary reorganization of the Jardin du Roi and its conversion into the Muséum d'histoire naturelle, Lamarck was given charge not of the botanical section, but of the section dealing with 'insects and worms'. This meant in practice what we would now call invertebrate zoology, an area in which Lamarck

Fig. 5.1 Jean-Baptiste Lamarck (1744–1829) in his blind and impoverished old age

had no particular expertise except that he had long been inter-ested in molluscs and had assembled an impressive collection of shells. The new appointment was nonetheless, from a historical point of view, a remarkable and unexpected success, since it was the study of invertebrates that led Lamarck to his theory of evolution. The similarities that he saw between some fossil mol-luscs and living specimens could hardly fail to suggest that there was a close affinity between the two. Moreover, the differences between related species were in some cases so slight that an extensive ordered series could easily be constructed. The con-clusion that Lamarck drew from these observations was that spe-cies, once created, were not fixed for all time, as was generally assumed, but that, on the contrary, they underwent continuous transformation, which took place very gradually and over immense periods of time.[3] There were, of course, tentative evolu-tionists before Lamarck, but none before him had advanced a consistent theory that was based on a substantial body of detailed observation. In the English-speaking world Lamarck is remembered more for his monumental errors than for his monumental contributions. True, he postulated that the observed changes in species were mediated by the inheritance of acquired traits; he regarded evolution as an orderly progression from the most primitive forms of life to the most complex (the *scala naturae*); he made many, and sometimes gross, anatomical and taxonomic errors. But the fact remains that if the theory of evolution can be attributed to one man, that man is Lamarck.

Lamarck's evolutionary ideas were not generally accepted, neither in his lifetime nor for several decades after his death. This public rejection was largely due to the anatomical skill, the academic ambition, and the imperious nature of Georges Cuvier (1769–1832) (Fig. 5.2). Cuvier was born in Montbéliard in the Duchy of Württemberg, the second son of a French-speaking

Fig. 5.2 Georges Cuvier (1769–1832)

family that had been Protestant for generations. He was christened Jean-Léopold-Nicolas-Frédéric, but after the death of his elder brother, Georges, he adopted that name and was thereafter always known by it. He was educated at the Karlsschule, a strongly evangelical school founded by the Duke of Württemberg, and chose studies that prepared him for a part in the administration of the state; but, finding no opening, he accepted a position as tutor to the son of a Protestant noble family in Normandy. Here his deep interest in natural history came to fruition, and the excellence of his dissections and illustrations drew him to the attention of l'abbé Tessier, a well-known agriculturalist, who recommended him to the scientific community in Paris. Through the good offices of Geoffroy St. Hilaire (Etiénne Geoffroy St. Hilaire, 1772–1844) he was offered an assistant professorship at the Jardin du Roi, and, after the institution's transformation into the Muséum d'histoire naturelle, he became the professor of comparative anatomy there.

Cuvier's public career was meteoric. Following the Restoration he was appointed chancellor of the University of Paris and a member of the cabinet of Louis XVIII; under Louis Philippe he was made a peer of France and Minister of the Interior. Despite his high office he retained the stern attitudes of evangelical Protestantism and was vitriolic in his dislike of the Catholic Church and its priesthood. Cuvier was unquestionably the most eminent comparative anatomist of his day and one of the founding fathers, if not the founding father, of the science of palaeontology. His voluminous publications on many different aspects of comparative anatomy remain a monument, and his study *The fossilized bones of quadrupeds*[4] inaugurated a field of investigation that is vigorously pursued to the present day.

Although Cuvier did publish a work on the anatomy of molluscs, the structure of invertebrates was not central to his research. It is perhaps for this reason that he failed to see the graded series of morphological changes that were obvious to Lamarck. Indeed, as he examined the fossil record, Cuvier was impressed by its discontinuities. In his view there was nothing about the structure of the surviving fossils to suggest that one species could be transformed into another. On the contrary, what he saw was the sudden extinction of one species and its replacement at a later period by a different species altogether. This extinction he thought to be the result of some immense natural catastrophe, an interpretation that we now call 'catastrophism'. For Cuvier the morphology of species was fixed: each retained indefinitely the characteristics with which it was endowed when it was created, and any subsequent modifications that it might undergo were minor and secondary. But this left Cuvier with no explanation for the appearance of new species after each catastrophe. On this matter, Cuvier declined to offer a clear opinion. Indeed, he took the view that such matters were beyond the scope of human enquiry, but it remains obvious from the whole tenor of his writings that he remained faithful to the belief that the creation of life, no matter how often it occurred, was always the product of divine intervention.

Lamarck would have none of this. For him the extinctions that Cuvier described were simply gaps in the fossil record, due perhaps in large part to the fact that so much of the world had yet to be explored; and being a mechanist not a vitalist, he sought explanations of the origin of life in some process other than divine intervention. Like all others who shared his assumptions about the nature of the material world, he adopted a version of spontaneous generation; but he believed that it involved only the smallest and most primitive forms of life. He could not conceive that more complex organisms, to say nothing of the gigantic creatures revealed by fossils, could spring to life completely formed. He envisaged that all 'higher' forms had developed originally from the elementary products of spontaneous generation, but only over exceedingly long periods of time. Lamarck was perhaps the first to appreciate how much time such a process might require. Cuvier had no idea. Because he found that the skeleton of a mummified dog was not significantly different from that of an eighteenth-century dog, Cuvier concluded that the morphology of the animal was fixed and that the evolution that Lamarck proposed was simply a myth.

In his *Natural history of invertebrates*[5] Lamarck gives a detailed description of how he envisages the mechanism of spontaneous generation. He proposes that in waters and humid places there are small, gelatinous, transparent particles that come together and form larger bodies. These attract fluid into their interior and thus form cavities that are encased in walls derived from the more viscous components of the gelatinous material. By the accretion of 'subtle, expansive fluids' a state of erethism is induced and this finally produces an 'orgasm'. It is the 'orgasm' that confers life on the inanimate material. All this, is, of course, pure fantasy, and Cuvier dismissed it as such. But he did not limit his attacks on Lamarck to speculations that exceeded the evidence. On numerous occasions he attacked the evidence itself, for, in arguing his case, Lamarck made more than one error of fact. In the end, Cuvier was completely triumphant. To his own satisfaction and that of the great majority of French scientists, he had demolished Lamarck's thesis and, by the exercise of academic and political authority, he had demolished the man himself. For several decades after both men had gone, evolution was not a live issue in France. That is perhaps why Darwin's ideas were so slow to take root there. But there were a few naturalists, even at the Muséum d'histoire naturelle, who found it difficult to accept Cuvier's view that the morphology of each species was fixed for all time. Geoffroy St. Hilaire, for example, whose protégé the young Cuvier had once been, found it difficult to see why a new species should be created harbouring the vestiges of organs that had served an obvious function in earlier species but that now served none. In reservations of this kind, the embers of the theory of evolution continued to flicker, but it was a long time before that fire was lit again.

In England religious orthodoxy prevailed. Charles Lyell (1797–1875), whose *Principles of geology* (1830–3) dominated geological thought for a generation, argued that the forces that operated in geological time were the same as those that still operated, and that these forces were natural, not supernatural. He rejected catastrophism and thought that change, except on rare occasions, occurred gradually. But he was a devout Christian, adhered firmly to the doctrine that life originated in an act of divine intervention, and believed that, although such intervention was possible at any time, it occurred rarely. Even Charles Darwin, whose theory of natural selection was difficult to reconcile with repeated supernatural intervention, did not explicitly deny that life on earth might have had a divine origin.[6] It is therefore not surprising that spontaneous generation, as an alternative to divine intervention, received short shrift in England.

From the beginning there was a marked hostility to the scheme proposed by Descartes. George Garden (1649–1733), a Scottish cleric who was at the same time a professor at King's College, Aberdeen, took the view that all the laws of motion failed to provide a convincing account of how plants or animals were formed. 'See', he wrote, 'how wretchedly Descartes came off.'[7] And William Harvey, although he believed in spontaneous generation, rejected out of hand any mechanism resembling that proposed by Descartes. There is little evidence that the experiments of Needham and Buffon found much resonance in England in the eighteenth century. The visionary physician Erasmus Darwin (1731–1802) did indeed favour the idea that life was generated, not created; and he conjectured that it sprang from microscopic forms.[8] But Erasmus Darwin's views were undermined by the work (and polemic) of Joseph Priestley (1733–1804). Like Spallanzani, Priestley showed, by the use of closed, partially closed, and open vessels, that the growth of silk-weeds in organic infusions was not due to spontaneous generation but, in all probability, to air-borne contamination.[9] It was not until the second half of the nineteenth century that Needham's experimental system was resurrected and developed, with remarkable intransigence by Bastian, of whom more later. In Germany, the idea of spontaneous generation was given a more sympathetic reception.

Notes

1. R. Descartes (1897–1913), 'Formation de l'animal', *Oeuvres* (ed. C. Adam and P. Tannery) **11**: 277. Paris.
2. D. Diderot (1951), *Le rêve de d'Alembert*. Marcel Didier, Paris.
3. J.-B. Lamarck (1797), *Mémoires de physique et d'histoire naturelle*. Paris (1809), *Philosophie zoologique*. Paris.
4. G. Cuvier (1812), *Recherches sur les ossemens fossiles des quadrupèdes*. Paris.
5. J.-B. Lamarck (1815–22), *Histoire naturelle des animaux sans vertèbres*. Paris.
6. C. Darwin (1859), *On the origin of species by means of natural selection*. London.
7. G. Garden (1691), *Phil. Trans. Roy. Soc.* **16**: 476.
8. E. Darwin (1794–6), *Zoonomia, or the laws of organic life*. London.
9. J. Priestley (1809), *Trans. Amer. Phil. Soc.* **6**: 119.

CHAPTER 6

Spoilt air

The advent of the nineteenth century brought with it a philosophical movement that quickly acquired and still bears the
name *Naturphilosophie*. It was a selective amalgam of the ideas of
Fichte, Hegel, and several other German idealist philosophers,
and was consolidated as a discipline in its own right principally
by Friedrich Wilhelm Schelling (1775–1854), whose *Introduction
to a draft for a system of Naturphilosophie*[1] appeared in 1799. This
was a system that never commended itself to empiricists. It aimed
not to extract information piecemeal from observations, but to
construct general principles that would embrace the whole of the
natural world and, at the same time, link it to the supernatural.
Based, for the most part, on a Spinozistic kind of pantheism,
these general principles seldom took into account the vagaries of
empirical evidence. *Naturphilosophie* was immensely attractive
because it provided an unimpeded answer to everything; and
for the best part of half a century it dominated the philosophy
of science in Germany.

One of the features of *Naturphilosophie* was that it sought to
construct principles that would not only embrace all forms of life,
but would also link the animate to the inanimate. The natural
bridge for such a transition was, of course, spontaneous generation. Hermann von Helmholtz (1821–94), one of the great nineteenth-century figures of both physiology and physics, insisted
repeatedly that the laws of physics must also apply to biology; and
the laws of physics did not include the repeated intervention of
the supernatural. Ernst Haeckel (1834–1919), whose voluminous
zoological and semi-philosophical works were widely read in
Germany, took spontaneous generation for granted. He was a
strong advocate of evolution and propagated a monistic view
of the world in which animate and inanimate matter were a
continuum governed by the same laws. Most *Naturphilosophers*

accepted that inanimate matter could come to life through the agency of completely natural mechanisms. But there were those who disagreed. Oken (1779–1851), whose *Textbook of Naturphilosophie*[2] ran to three editions, argued vehemently that all living things came from other living things (*omne vivum e vivo*). Ehrenberg, whose work on the infusorians has already been mentioned,[3] assembled a convincing body of evidence against the doctrine that these minute organisms were produced by spontaneous generation; their anatomical structure was too complex, and no intermediate forms could be seen.[4] Johannes Müller (1801–58), the most influential German physiologist of his day, took the view that the random assembly of inanimate particles could not possibly generate organic matter.

The advocates of spontaneous generation had, moreover, to face the experimental evidence produced by Spallanzani. This they did, almost in unison, by asserting that his experimental procedure was faulty and, notably, that the heat he applied had in some way spoilt the air in his flasks and rendered it incapable of supporting the generation of life. This was the argument advanced, for example, by Gottfried Reinhold Treviranus (1776–1837), a practising physician in Bremen, who, mainly through his botanical work, acquired a formidable European reputation as a naturalist. Treviranus, much influenced by the version of *Naturphilosophie* that he had adopted, asserted that in Spallanzani's experiments there was either too little air to permit the development of microorganisms, or that the air had been spoilt by heating.[5] These objections were reinforced by the observations of the eminent French physicist and chemist Louis Joseph Gay-Lussac (1778–1850), who in 1810 showed that the air in vessels that had been heated and hermetically sealed contained no oxygen, which, in his view, was essential for the decomposition of organic materials.[6]

The problem confronting experimentalists working with flasks was therefore twofold: to shield the infusion from contamination by organisms that might be present in the surrounding air, but at the same time to ensure that the supply of air was adequate; and to purify the air by some method that did not involve heating it. Although 'spoilt air' or 'not enough air' remained the principal objections to Spallanzani's experiments for decades, it was not until 1836, more than seventy years after the publication of his criticisms of Needham and Buffon, that an attempt was made to overcome these objections. In that year Franz Schulze (1815–73), a chemist in Berlin, published the results of his findings with a new type of apparatus designed to purify the air by passage through sulphuric acid (oil of vitriol).

The paper appeared in the widely read Annalen für Physik[7] and was described as a preliminary investigation; but no further paper by Schulze concerning spontaneous generation appears to have been published. It begins with the interesting statement that it has long been agreed that boiled infusions in hermetically sealed vessels did not generate microorganisms. Since Spallanzani had himself found that some animalcules or their 'seeds' resisted boiling, Schulze's generalization obviously overlooks a good deal of evidence to the contrary. He then turns his attention to the requirement for air. He accepts that the growth of microorganisms requires an adequate supply of air, and mentions, in support of this view, that if the infusion is covered with a layer of oil, no growth occurs.

The apparatus that Schulze assembled to avoid the use of heat and to ensure an adequate supply of air is shown in Fig. 6.1. This

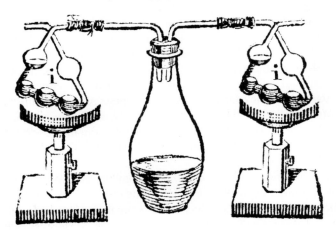

Fig. 6.1 Schulze's drawing of the apparatus he designed to sterilize the air entering the flask. On one side there is a trap containing sulphuric acid; on the other a trap containing caustic potash. Air can enter the flask only via one or other trap (see text)

can be reproduced here because few of the early experimenters who explored the question of spontaneous generation actually illustrated the apparatus that they used; in most cases the exact nature of the equipment has to be construed from the written word. The flask at the centre of the diagram was half-filled with distilled water to which various animal and vegetable materials were added. It was sealed with a 'good cork' through which two holes had been drilled to accommodate two glass tubes. The cork and the tubes that traversed it were said to form an air-tight seal. Each of these tubes was connected to a container composed of five spherical traps, the two outside ones being slightly larger than the inner three. The flask containing the infusion of organic material was placed in a heated sandbath until the infusion boiled vigorously, and while the steam was still issuing from the two tubes, concentrated sulphuric acid was introduced into one set of traps and a solution of caustic potash into the other. Air could thus enter the flask only by passing through one or other of the corrosive fluids. Schulze assumed that all living organisms within the flask or the connecting tubes had been killed by the heat of the sandbath.

The apparatus was set beside a window during the summer months in order to supply both warmth and light, which some proponents of spontaneous generation declared were necessary. Several times each day fresh air was aspirated into the flask by mouth; the air in the flask was gently withdrawn via the caustic potash traps and replaced by air that entered the flask through the sulphuric acid traps. Schulze argues that 'the composition of the air was, of course, not altered by passage through the sulphuric acid', but that all living, or potentially living, things must be destroyed. The controls were flasks containing the same infusions treated in the same way, but exposed to the surrounding air.

The experiments were continued from the end of May until the beginning of August, but at no time did the infusions in the shielded flasks show any life, either animal or plant. When the apparatus was finally dismantled there was still no trace of 'infusoria, silk-weeds or moulds'. On the other hand, the infusions in the flasks that had been left open teemed with all three within a few days. In fact, on the very first day, the open flasks contained 'vibrios' and 'monads', followed a little later by 'polygastric infusoria' and rotifers (complex protozoa).

These were the experiments of an organic chemist rather than a biologist, but they were very powerful. They did not, however, change the minds of the committed. Apart from the usual litany of preconceived opinions, there were two technical objections to Schulze's experiments. The first was that the bubbles of air passing through the sulphuric acid might not have been sterilized, so that living organisms within the bubbles, and hence protected from the corrosive fluid, might nonetheless have been introduced into the flask. While this was obviously a serious objection to the use of any liquid trap in experiments of this kind, it was not a valid criticism of Schulze's experiments; for, in his hands, the infusions that received air only via the sulphuric acid did not produce any microorganisms. The second objection was again that the incoming air might have been spoilt by the acid. Schulze's chemist's view was that the composition of the air would not, of course ('natürlich'), be changed by passage through sulphuric acid, but it remained possible that vapour from the sulphuric acid or volatile impurities in it might have been carried over into the flask and thus inhibited the growth of microorganisms.

In any case, Schulze's unblemished result must, to some extent, be regarded as fortunate, for Spallanzani had already provided ample evidence that there were microorganisms that survived the temperature of boiling water. Schulze, however, did not use temperatures higher than this to ensure the sterility of his materials: the infusion was merely boiled and the glassware exposed to the steam thus generated. It is a pity that Schulze's 'preliminary communication' was not followed by a more detailed account of the work. It would have been of interest to know how many flasks were examined, exactly what it was that he put into the infusions, and whether there was any change in the acidity of the infusions after they had received air via the sulphuric acid trap. Why there was no definitive paper remains obscure. It is possible that Schulze regarded the matter as settled; or that his interests simply moved elsewhere; or, perhaps more probably, that he was discouraged by the weight of ciriticism that his work received.

A year later, in 1837, another 'Preliminary report' appeared,[8] this time by Theodor Schwann (1810–82) (Fig. 6.2), whose

Fig. 6.2 Theodor Schwann (1810–82)

contribution to our understanding of the significance of cells has been described in detail elsewhere.[9] His work on spontaneous generation antedates his observations on animal and plant cells[10] and is, in many ways, more interesting, certainly more rigorous. It remains surprising that the meticulous experiments of 1837 should within a couple of years have given way to the wildest of assumptions about the mechanism of cell formation.

Schwann's interest in spontaneous generation must have been awakened very early in his career, for, in his university lecture notes, he has a quotation from Johannes Müller to the effect that the discovery of sexual organs in many infusorians threw doubt on the theory of spontaneous generation. Moreover, in the second edition of his famous *Textbook of human physiology*,[11] Müller states that 'those who have examined the effects of atmospheric air on boiled organic material cannot prove that the infusoria or moulds produced do not arise from dried infusoria or their seeds brought in with the dust in the atmospheric air'. Since Schwann worked in Berlin as Müller's *Assistent* from 1834 to 1839, it seems likely that it was Müller who suggested that spontaneous generation might be a suitable subject to investigate, or, at least, encouraged the investigation.

Schwann's initial experiments on this subject were not published by Schwann himself, but accounts of them appear in the official report of the meeting of German naturalists and doctors in Jena in September 1836[12] and also in Lorenz Oken's abstracting journal, *Isis*.[13] In the first of these there is no more than a brief summary of Schwann's communication. He apparently demonstrated a glass vessel containing a small amount of an organic infusion but largely filled with air. This had been boiled and sealed hermetically by fusing the neck. He had not yet observed any infusoria in it and is said to have concluded that spontaneous generation did not therefore exist. As described, Schwann's experiments appear to be no different from those of Spallanzani and his followers. Schwann was neither generous nor meticulous in his references to the work of his predecessors, but it is difficult to accept that he did not know of the work of Spallanzani and the controversy that it provoked. In any case, the official report of the meeting mentions that several distinguished professors in the audience doubted whether the experiments that Schwann described warranted the conclusion that he drew from them.

One of the objections raised at the meeting was that boiling the organic mixture might have converted the oxygen in the air to carbon dioxide, which was held to be toxic. To answer this objection, Schwann made some important modifications to his procedure. The original experiment and the modifications are described more fully in Oken's journal. Here we are first given some information about the composition of the infusion and how it was heated. The infusion was made with muscle tissue and

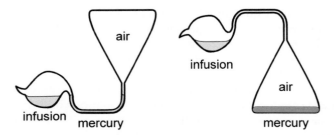

Fig. 6.3 Reconstruction of the apparatus used by Schwann in his earlier experiments. The vessel containing the infusion is separated from the air in the flask by a barrier of mercury. When the apparatus is inverted, the mercury flows to the bottom of the flask and air now has access to the infusion (see text)

was immersed for half an hour in boiling water or in a 'Papin's digester', in which the boiling water is under pressure. Schwann apparently assumed that the temperature of the infusion must therefore have reached 80 °C (Réaumur) (the boiling point of water), but he either did not know or took no cognizance of Spallanzani's observations on 'seeds' that were resistant to this temperature. To test the idea that the continued sterility of his boiled infusions was due to carbon dioxide toxicity, Schwann devised an apparatus that was much more complicated than a sealed flask. A representation of this apparatus, construed from the text, is shown in Fig. 6.3.

The most important innovation, and one heavily criticized later by Pasteur, was the introduction of mercury as a barrier between the boiled infusion and the supply of air. The neck of a large flask was drawn out into a loop which was expanded at one end to form a small chamber to hold the infusion. Into this loop enough mercury was deposited to make an effective barrier when the flask was inverted. The infusion was introduced into the small chamber, which was sealed hermetically by fusing the glass aperture. The whole apparatus was heated to the temperature of boiling water as before and then restored to the upright position. This was done in a manner that deposited the mercury to the bottom of the flask and thus permitted the air that it contained to come into contact with the boiled infusion. Any carbon dioxide formed by boiling the infusion in the small chamber would thus have been diluted by a large volume of air. As before, Schwann found no growth of microorganisms in the infusion and concluded once again that there was no evidence to support the theory of spontaneous generation. Müller thought Schwann's experiments so important that he gave an account of them to the February 1837 meeting of the Berlin Society of Friends of the Natural Sciences (*naturforschende Freunde*).

Schwann's early experiments did not, however, meet the objection that heating might in some way have deprived the air of the vital force that was necessary to produce life; nor did they eliminate the possibility that a continuous flow of air was necessary. Schwann was aware of Schulze's experiments and recognized the difficulties inherent in the use of fluids to sterilize the air. He was therefore obliged to rely on heat, but he tested the capacity of the heated air to support life, which had not previously been done.

In his 'Preliminary report' of 1837[14] Schwann begins by summarizing the experiments that he reported in Jena the previous year. He emphasizes the fact that the sealed flask that he used contained only a little of the muscle infusion and a lot of air, so that it was unlikely that the failure of the boiled infusion to produce any microorganisms was due to exhaustion of oxygen. Nonetheless, he conceded that a continuous flow of air was desirable and, to achieve this, he devised the following experiment. A flask containing a little muscle meat and water to about one-third of its volume was stoppered with a cork perforated by two glass tubes. Before communicating with the surrounding air, each of these tubes passed through 3 inches (7.5 centimetres) of a slightly fluid metal mixture maintained continuously at the boiling point of mercury. One of the tubes was connected to a gasometer (a container in which the gas, in this case air, is stored and can be delivered at the pressure desired). The flask and its contents were then heated to the boiling point of water. For several weeks after that air from the gasometer was passed continuously through the flask, entering via one tube and leaving by the other, but entering only after transit through the hot metal bath. No putrefaction of the infusion was seen nor any growth of micororganisms within it. However, in a footnote, Schwann explains that this arrangement has the serious disadvantage that it is very difficult to maintain the required temperature in the hot

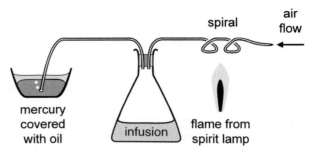

Fig. 6.4 Reconstruction of the apparatus used by Schwann in his later experiments. Air can enter the flask only via the heated glass spiral (see text)

metal bath over long periods of time. He therefore makes one further modification in the equipment. It is the last modification that his published work records; a reconstruction of it, again from the text, is shown in Fig. 6.4.

The essential change is the substitution of a heated glass coil, instead of a hot metal bath, through which the incoming air must pass before entering the flask. This has a volume of 3 ounces and is quarter-filled with water and muscle tissue. It is then stoppered with a cork through which pass two glass tubes as before. The cork is held down with wire and covered with a thick solution of rubber in linseed oil diluted with turpentine. One of the glass tubes is drawn out into a coil with an outlet to the surrounding air. The other passes to a bowl of mercury covered with oil. A small amount of corrosive sublimate is deposited on the surface of the mercury to kill off anything that might grow in the water condensing between the mercury and the layer of oil. The flask is heated until steam issues from the tubes and the oil-covered mercury is hot enough to prevent the steam condensing on its surface. A spirit lamp, placed beneath the glass coil, is adjusted to provide enough heat to soften the glass. Water droplets condensing on the cooler parts of the coil are driven off with a second spirit lamp. The flask is again heated, this time for a quarter of an hour, and air allowed to enter, but only via the heated coil. The aperture of this coil is then drawn out to a fine point and sealed by fusing the glass. The air trapped between the coil and the seal is heated especially thoroughly (*besonders ausgeglüht*), and the spirit lamp then removed.

This procedure is repeated from time to time. The coil is heated, the sealed tip of the tube broken off, and the air slowly introduced into the flask via the heated coil. The tip of the tube is then re-sealed, and the section between the tip and the coil again thoroughly heated. Thereafter, the flask is maintained at a temperature between 14° and 20° Réaumur for up to six weeks, but still there is no putrefaction and no growth of microorganisms. But both occur within a few days when the flask is opened. Finally, Schwann sought to meet the objection that the air was somehow spoilt by the heat. This he did by putting a frog inside the flask and showing that it survived in the pre-heated air. No details are given of this experiment, but it at once attracted the objection that while the air might support life, it had lost the power to generate it.

I have described these experiments in some detail to illustrate two points. The first is to show how difficult it then was to devise an apparatus that could deliver a regular supply of sterilized air to the flask; and the second is to show just how meticulous an experimenter Schwann was. Even so, with all the precautions that he took, he seems initially to have found putrefaction in some of his heated flasks, for he stresses that great care must be taken to ensure that the cork stopper and the glass tubes fit tightly. His conclusion is very circumspect: 'For those who oppose *generatio aequivoca*, these results would appear to indicate that the seeds or germs of the moulds and infusoria, which according to this view are present in atmospheric air, are destroyed by intensive heating of the air.' The excessive modesty of this conclusion is, no doubt, a reaction to the criticisms that had been made of his previous experiments. It is also of interest, in the light of Schwann's later ambiguities, that he does not now openly admit that he is one of 'those who oppose *generatio aequivoca*'. I shall have more to say at a later stage about Schwann's vacillation.

The rest of Schwann's 'Preliminary report' is largely concerned with the fermentation of wine. The equipment used is not described in detail but is evidently a variant of that described above. Four small flasks are filled completely with a solution of cane sugar and beer yeast, immersed in boiling water for 10 minutes, and, while still hot, inverted under a layer of mercury. When the flasks have cooled, one-quarter to one-third of the solution in them is displaced by air introduced via a coil as before. In one pair of flasks the coil is heated; in the other pair it is not. The flasks are then corked and maintained for 4–6 weeks at 10°– 14° Réaumur. The solution fermented in the flasks containing air that had not been heated, but did not ferment in the flasks containing pre-heated air. Schwann goes on to analyse the composition of the air after it has passed through the heated coil. He finds that the oxygen concentration is still 19.4 per cent, not far from the concentration found in atmospheric air. It is thus clear that lack of oxygen cannot be the limiting factor when fermentation does not occur.

The results described by Schwann in the main part of the text are unequivocal, but he admits that in later experiments the results were inconsistent. In a footnote he explains why this might have been so. He does not envisage the possibility that the inconsistent results might also reflect genuine biological phenomena, but regards the inconsistencies as no more than a consequence of faulty procedures. To rectify these, further elaboration of the equipment would, in his view, be required. The fact that he treated the counter-examples as artefacts indicates that in the matter of putrefaction and spontaneous generation Schwann was not neutral. Nor was anyone else who undertook to investigate the problem. But the variability of his later results might explain why the 'Preliminary report' was not followed by a definitive paper, although one was apparently planned. In any case, this variability, like that described by Spallanzani, was enough to induce those who accepted the reality of spontaneous generation to regard Schwann's experiments as inconclusive.

When Schwann looked at his fermented solutions under the microscope, he noticed that they contained many 'round, but for the most part oval, granules, yellowish white in colour and growing in rows'. By the use of reagents that were thought to be selectively toxic, he reached the conclusion that these granules were plant forms of some kind. He had them examined by Professor Franz J. F. Meyer (1804–40), an eminent botanist at the University of Berlin, who considered that they were algae or organisms of the silk-weed family. Meyer favoured the latter because the granules contained no green pigment. Schwann found that beer yeast consisted almost entirely of such bodies, which grew visibly under the microscope, formed buds, and multiplied as fermentation proceeded.

Schwann's discovery of yeast cells was, however, anticipated by Cagniard de la Tour (1777–1859). Always sensitive to the question of priority, Schwann points out that the unaltered text of his paper was presented by Müller to the Society of Friends of the Natural Sciences in Berlin in the early days of February 1837. But similar observations on the fermentation of beer had already been communicated by Cagniard to the journal of the Institut in Paris and had appeared in an issue of that journal dated 23 November 1836. Schwann claims, and there is no reason to doubt it, that he did not receive the relevant issue of the French journal until after Müller's presentation in Berlin. But Schwann could only have made this comment in order to establish the independence of his own work, which was not in doubt. Actually, Cagniard may have anticipated him more decisively than the narrowness of the interval between their two publications might suggest, for Cagniard had also described yeast cells in fermenting extracts of other organic substances such as fruit or mash. In any case, in later publications, Schwann does not refer to the discovery of yeast cells by Cagniard de la Tour, but their discovery 'by myself and Cagniard de la Tour'.

The conclusions that Schwann finally drew from his experiments on fermentation and putrefaction were first, that neither of these processes, nor the production of the animal and plant forms associated with them, occurs if the organic infusion is heated, unless unheated air is admitted; and second, that fermentation and the formation of gas run *pari passu* with the growth of the microscopic bodies that he had discovered.

Schwann speculated on the role that his microscopic bodies, for which he proposed the name 'sugar-fungus' ('Zuckerpilz'), might play in converting the sugar to alcohol. These speculations were the starting point of a controversy that raged until the beginning of the twentieth century. On the one hand, there were those, like Claude Bernard (1813–78), the doyen of nineteenth-century French physiologists, who believed that such a simple chemical reaction should not require the presence of living organisms; and, on the other, those like Louis Pasteur who regarded living organisms as essential. The issue was not settled until 1897 when Eduard Buchner (1860–1917) extracted from pulverized yeast a cell-free preparation that underwent vigorous fermentation.[15]

Notes

1. F. W. Schelling (1799), *Einleitung zu einem Entwurf eines Systems der Naturphilosophie.* Jena and Leipzig.
2. L. Oken (1809), *Lehrbuch der Naturphilosophie.* Jena.
3. See Chapter 3, note 8.
4. C. G. Ehrenberg (1832), *Poggendorfs Annalen* **24**: 1.
5. G. R. Treviranus (1805), *Biologie oder Philosophie der lebenden Natur.* Göttingen.
6. L. J. Gay-Lussac (1810), *Annales de chimie* **76**: 245.
7. F. Schulze (1836), *Poggendorfs Annalen* **39**: 487.
8. T. Schwann (1837), *Poggendorfs Annalen* **41**: 184.
9. See Chapter 2, note 4.
10. T. Schwann (1839), *Mikroskopische Untersuchungen über die Uebereinstimmung in der Struktur und dem Wachstum der Thiere und Pflanzen.* Berlin.
11. J. Müller (1833–8), *Handbuch der Physiologie des Menschen.* Coblenz.
12. T. Schwann (1837), *Amtlicher Bericht über die Versammlung deutscher Naturforscher und Ärzte zu Jena in September 1836.* Weimar.
13. *Isis* **7** (1837): col. 524.
14. See note 8.
15. E. Buchner (1897), *Ber. deutsch. chem. Gesell.* **30**: 117.

CHAPTER 7

Internal parasites

After Redi's experiments with insects, few believed that visible creatures were normally produced by spontaneous generation. As discussed earlier, the passage of centuries was accompanied by a progressive reduction in the size of the animals that were thought to be generated in this way. But whenever, in the course of this slow attrition of belief, the mode of reproduction of any animal remained obscure, there were those who argued that it was produced by spontaneous generation. The most notable example of this was the protracted debate about the origin of internal parasites, and, in particular, intestinal worms. Redi's famous book *Observations on living animals that are found within the bodies of living animals*[1] contains no empirical observations on their mode of reproduction, although, as mentioned previously, he did envisage some form of complex life-cycle. But he did not see anything implausible in the idea that the parasites were generated by the tissues of their hosts. And, in the absence of evidence to the contrary, this idea persisted well into the nineteenth century.

Ehrenberg, in the paper already mentioned,[2] discusses intestinal worms in some detail. Again, he points out that they have an elaborate internal structure, including an egg-laying apparatus, and that they produce huge numbers of eggs. He finds it difficult to believe that such creatures could be generated by spontaneous generation. He does not hazard a complete life-cycle, but proposes that the eggs, released in large numbers into the bloodstream, develop only at sites where the environment is propitious. To a modern eye Ehrenberg's arguments against the spontaneous generation of intestinal worms seem convincing enough, but they did not convince many of his contemporaries. Carl Friedrich Burdach (1776–1847), professor of anatomy in Dorpat and later in Königsberg, whose *Die Physiologie als Erfahrungswissenschaft*

(Physiology as an empirical science) (1826–40) is shot through with *Naturphilosophie*, argued that those who supported the theory of spontaneous generation did so on the basis of genuine experimental evidence. Johann Georg Bremser (d. 1827), curator at the natural history museum in Vienna, who made a thorough study of the worms that infest man, concluded that of the four possible ways in which these parasites might be generated, spontaneous generation was the most likely. And Félix Dujardin (1801–60), who appears to have been the first to describe what we now call protoplasm, was still arguing in favour of the spontaneous generation of worms in 1845, more than a decade after the appearance of Ehrenberg's paper. On the other hand, C. T. von Siebold (1804–65), probably the leading German invertebrate zoologist of his day, did not believe in spontaneous generation, but he agreed with Dujardin that the parasites found in intermediate hosts were to be regarded as aberrant forms that had strayed away from their natural environment. This dissension was resolved by the work of three remarkable men: Adelbert von Chamisso (1781–1838), Johannes Japetus Smith Steenstrup (1813–97), and Friedrich Küchenmeister (1821–90).

Chamisso (Fig. 7.1) was a man whose place in literature is no less assured than his place in science. His full name was Louis Charles Adélaïde de Chamisso, Vicomte de Boncourt. His was an aristocratic family whose ancestral home was the Château de Boncourt in the Champagne. During the Revolution, when Adélaïde was 19, the family abandoned France and settled in Prussia, which rapidly became a haven for French emigrés. The Adélaïde was converted to Adelbert (not the more commonly spelt Adalbert) and the de Chamisso de Boncourt shortened to the Germanic von Chamisso. It is as Adelbert von Chamisso that he has gone down in history. By some accounts he is said nonetheless to have retained a slight French accent. In 1797 he became a page to Queen Friederike Luise of Prussia and from 1801 to 1806 served as an ensign in the Prussian Army. It was during his army service that he acquired an interest in philosophy and literature, but his interest in science was not awakened until he returned briefly to Paris in 1806 and attended lectures at the Collège de France. Not finding post-revolutionary France to his liking, he returned to Prussia where he decided to further his scientific education by

Fig. 7.1 Adelbert von Chamisso (1781–1838)

reading natural sciences at the University of Berlin. Science did not, however, completely displace literature, and in 1814 he produced the novella *Peter Schlemihls wundersame Geschichte* (The wondrous tale of Peter Schlemihl). The book was an instant success and has remained one of the classics of Romantic German literature.

It is the story of a young man who sells his shadow in exchange for a purse that provides inexhaustible wealth. But, without a shadow, he is unable to enjoy his wealth and even loses the woman he loves. He is, however, unwilling to sell his soul to regain his shadow and struggles on until he discovers a pair of seven-league boots. These at last give him the power to devote himself to scientific research. It is in research that he finds solace and some compensation for the loss of his shadow. The symbolism of the tale has been interpreted in a number of ways, but the autobiographical overtones are obvious.

A year after *Peter Schlemihl* appeared, an undecided Chamisso managed to obtain a post as a botanist aboard the *Rurik*, a Russian brig that was about to undertake a worldwide voyage of exploration. The voyage lasted three years, and it was during this period that Chamisso, in addition to many botanical observations, made his great discovery—the phenomenon that we now call 'alternation of generations'. It was then an axiom of biology that 'like begets like', but Chamisso found that in salps (a kind of sea-squirt), like did not beget like. The salps begat progeny that did not at all resemble their parents either in appearance or in behaviour; but when that first generation of radically different creatures itself begat progeny, these again assumed the morphology not of their parents, but of their grandparents. In this way the generations alternated, but the stability of the species was nonetheless maintained.

When, a year after the expedition was over, Chamisso published his *De Salpa* (On salps),[3] the magnitude of the discovery was at once recognized by the biological community in Berlin. In quick succession, he was given an honorary doctorate of the University of Berlin, was made an honorary member of the Berlin Society of Friends of the Natural Sciences, and was elected into the Leopoldina, the foremost Prussian scientific academy. In the same year he was appointed assistant curator of the Royal Botanical Gardens, where his friend and collaborator D. F. L. von Schlechtendal was curator. When, in 1833, Schlechtendal left, Chamisso was made curator, and in this post he remained to the end of his life.

There has been some needless misunderstanding about who it was that discovered alternation of generations. Most parasitology texts attribute this to Steenstrup, and one or two less plausible candidates have put in their own claims. The matter is clear: Chamisso's *De Salpa* antedates Steenstrup's monograph on the alternation of generations[4] by 23 years. What distinguishes the two works is this. Chamisso's observations dealt with only one family of organisms, and although, as previously mentioned, the importance of his observations was at once acknowledged, it was not yet known whether alternation of generations was a peculiarity of salps or whether this form of life-cycle was more widely distributed in the animal kingdom. Steenstrup showed that it was and, most importantly, that it was the normal mode of reproduction of parasitic tape-worms, including liver-flukes. It was this observation that captured the attention of the many investigators still struggling to understand how internal parasites were generated; and it sounded the death-knell of the idea that they were generated spontaneously.

But the *De Salpa* is not merely the record of a chance observation. Chamisso examined, and described in detail, 11 different species of salps and fully understood what was going on. On the second page of his monograph he sets out his conclusions with admirable clarity. He writes:

> Salps are seen in two forms, one solitary and the other composed of an aggregate of contiguous organisms. The solitary organism generates multiple progeny, but the multiple aggregate generates only a single offspring. The two morphologies are stable throughout the life of the organism. The progeny of each form does not resemble its parent, whether it is androgynous [producing only male offspring] or simply feminine, but resembles its grandparents. That is to say, the mother does not resemble the daughter, but resembles her sisters, her grandparents and her grandchildren. There is thus a stable alternation of generations.

Chamisso was well aware of the implications of what he had discovered, for the *De Salpa* was the very first of the series of publications that issued from his work aboard the *Rurik*.

Two years before he succeeded Schlechtendal as curator of the Royal Botanical Gardens, Chamisso published a collection of his poems. These are, for the most part, slightly atypical examples of the Romantic German lyric, atypical because some of them, such as 'Die alte Waschfrau' ('The old washerwoman') or 'Der Bettler und sein Hund' ('The beggar and his dog'), display a realism that is characteristic of a later age. 'Das Schloss Boncourt' ('Boncourt Castle') is one poem that has found its way into numerous anthologies of German poetry. This is a nostalgic piece about the family home in the Champagne, the obliterated Château de Boncourt. It was first published as an appendix to the second edition of *Peter Schlemihl*, which makes it clear that the shadow that Chamisso himself had lost was the bond with his ancestral roots in France.

Johannes Steenstrup was a Dane who for many years held the chair of Zoology at the University of Copenhagen. His monograph on the alternation of generations was written in Danish[5] and first appeared in 1842. Despite the language barrier, it quickly made its way, and it was soon followed by a German version, which was supervised by Steenstrup himself. An English translation of the German version appeared in 1845.[6] The wave of enthusiasm that greeted Steenstrup's work was certainly justified, but it generated the impression that it was he who discovered the alternation of generations. Apart from Chamisso, whose scientific work seems to have been largely forgotten, except perhaps in Berlin, there were other precursors. Michael Sars (1805–69) was one. He was born in Bergen and died in Christiana (Oslo), where in 1854 he was appointed *Professor extraordinarius* (the nearest equivalent is adjunct professor). Prior to this, for a period of some 25 years, he served as a teacher, vicar, and later rector of the seashore communities in western Norway. During this period he acquired a notable reputation as a zoologist, and his observations on the alternation of generations in jellyfish provided the evidence on which much of the first section of Steenstrup's monograph was based. Observations of the same kind had been made by von Siebold and Karl von Baer (1792–1876), but none of these authors described a complete life-cycle or appeared to grasp the significance of such cycles. That achievement is Steenstrup's.

His monograph is divided into four sections. The first deals with jellyfish. Steenstrup describes how the motile jellyfish give rise to progeny that are sessile and, when attached to solid surfaces, resemble polyps. The second generation, however, assumes the observable features of the grandparent, not the parent. In the second section of the monograph, alternation of generations is again described, this time in claviform (club-shaped) polyps. These generate an intermediate generation of detached free-swimming organisms that resemble small jellyfish. Then there follows a section on salps which recapitulates the observations that had been made more than 20 years earlier by Chamisso. But the *pièce de résistance* is the last section, which describes the life-cycle of flat-worms, especially that of the liver-fluke. It is in this section that the idea of an intermediate generation is expanded to include that of an intermediate host. For the future of parasitology, this concept was crucial. Steenstrup demonstrates that the flukes produce eggs, which, at one stroke, undermined the notion that the flukes were generated spontaneously. The eggs hatch to produce free-swimming embryos, which then infest the intermediate host, a water-snail, or a mollusc. The free-swimming embryo gains access to the intermediate host by attaching itself to a layer of external mucus and then burrowing into the interior. There it forms a sac that contains a number of germ cells which give rise to the individuals of the next generation. These are also motile and have a sucker, intestine, and tail. When mature they escape from the intermediate host, remain free in the water, or attach to water plants. In some cases, they may infest a second intermediate host, a freshwater fish, or a crab. They thus find their way into the food chain of man and other animals, and there in due course produce fully developed flukes. Steenstrup concludes with the observation that there is an obvious resemblance between the alternation of generations in lower animals and the reproductive processes of plants. But, much to Darwin's chagrin, Steenstrup never accepted the theory of evolution by natural selection.

I have described the contents of this section of Steenstrup's monograph at some length to show how, in his hands, alternation of generations ceased to be a curiosity and became the principle that guided the development of a whole new science. But there was one step in the life-cycle of flat-worms that remained unsupported by direct experimental evidence. It was assumed, plausibly enough, that the intermediate generation that remained free in the water or attached to water plants, fish, and crabs were indeed the organisms that infested animals and gave rise to the mature worms. But no one had yet demonstrated that this was actually the case, and certainly not in man. That demonstration we owe to Küchenmeister.

It does not appear to have been Steenstrup's monograph that prompted Küchenmeister's experiments, but a religious fundamentalism that made it impossible for him to accept von Siebold's theory that the intermediate generations of worms were aberrant forms that had strayed from their main line of development. Küchenmeister was a fanatically religious man who believed not only that all living creatures were created by God, but also that they were created for a purpose. The anatomical features that they displayed were not therefore degenerate structures, but had been designed to fulfil the purpose for which each organism was created. Dujardin's view that intermediate forms were freaks ('une sorte de monstruosité')[7] was anathema to Küchenmeister who, of course, held that God made nothing in vain. It was the belief that intermediate forms must be a part of God's design that induced Küchenmeister to do the obvious thing—to feed the intermediates to appropriate hosts then see whether they did indeed give rise to the mature worms.[8]

Küchenmeister's experiments centred on the transformations of tape-worms. An intermediate larval stage in the development of the tape-worm is known as a bladder-worm because the whole of the larva forms a bladder into which the head that eventually gives rise to the mature worm protrudes. By feeding the bladder-worms from a variety of sources to potential hosts (dogs, cats, foxes), Küchenmeister showed beyond doubt that they were indeed the infective larval forms of mature tape-worms.[9] Küchenmeister's identification and classification of his material was sometimes faulty, but his experiments provided the impetus for similar ones from many other workers, and the essential conclusion that he drew was soon generally accepted. Finally, he extended his research to man. For this investigation he chose a delinquent in whom he grew various kinds of mature tape-worms from the bladder-worms, including the characteristic 'armed' tape-worms generated by larvae present in infected pork.

By the 1850s no one argued any longer that intestinal parasites were the products of spontaneous generation.

Notes

1. See Chapter 2, note 5.
2. See Chapter 6, note 4.
3. A. von Chamisso (1819), *De Salpa*. Berlin.
4. J. J. S. Steenstrup (1842), *Om Forplantning og Udrikling gjennem vexlende Generationsraekker*. Copenhagen.
5. Ibid.
6. J. J. S. Steenstrup (1845), *On the alternation of generations* (trans. from the German by George Busk). London.
7. F. Dujardin (1845), *Histoire naturelle des helminthes ou vers intestinaux*. Paris.
8. F. Küchenmeister (1853), *Zittau*, 10–12.
9. F. Küchenmeister (1857), *Z. klin. Med.* **2**: 240, 295. Leipzig.

CHAPTER 8

Cotton wool

Schwann's experiments, with the help of Müller's advocacy, were taken seriously, although they were not by any means universally accepted. Helmholtz, whose general philosophy inclined him to accept the idea that there was a natural transition from the inanimate to the animate, took them seriously enough to repeat them.[1] His experimental design was essentially the same as Schwann's, but the apparatus was a little simpler. A flask containing the organic infusions (grape juice, meat, etc.) is sealed with a cork through which pass two glass tubes. One is bent so as to provide a convenient sucking tube; the other is drawn out to a fine tapered end. The outlet of the flask with its perforated cork is covered with sealing wax. The infusion is boiled, and while it is still steaming the sucking tube is closed off. A lighted spirit lamp placed beneath the tapered end of the other tube ensures that the air entering the flask is first heated thoroughly. To provide an adequate supply of oxygen, the sucking tube is re-opened from time to time and air gently sucked into the flask via the orifice heated by the spirit lamp. Helmholtz confirms that no putrefaction of the infusion occurs if the air entering the flask is first heated, but putrefaction does occur if the incoming air is not heated. While these experiments added weight to Schwann's principal conclusion, they did not in any way meet the objection that the heat had somehow spoilt the air. That objection was met by the work of Heinrich Schröder (1810–85) and Theodor von Dusch (1824–90), whose essential contribution to the debate was that they filtered the air instead of heating it.

A twentieth-century microbiologist, for whom the use of tubes plugged with cotton wool is second nature, might have difficulty in understanding why some fifteen years elapsed between Schwann's use of a spirit lamp to sterilize the incoming air and Schröder's use of a cotton wool plug. The first and compelling reason is that modern cotton wool, the compressible absorbent material produced by extracting the wax from raw cotton, was not widely available until the middle of the nineteenth century, and then it was used mainly as wadding or packing material. The decisive step was not simply to use cotton wool, but to use it as a filter. This decision rested on two assumptions. The first was, as Schwann had suggested, that there was something particulate in the air that could be filtered out; and the second was that cotton wool might be an adequate filter. When, in 1854, Schröder and von Dusch published their paper,[2] the definitive account of Pasteur's studies on the particles in air[3] was still some years away.

The considerations that induced Schröder and von Dusch to try cotton wool as a filter are set out in the introduction to their paper. It seems to have been widely believed that woodland provided a screen against pestilential influences carried by the wind. Thus, a report by Rigaud de l'Isle is quoted in which he claims that the 'miasmatic influences of the Pontini marshes are warded off by nearby woods'. In a paper communicated to the Académie des sciences in 1853,[4] the same year as that in which Schröder and von Dusch submitted their paper, M. Becquerel gives an explicit account of what goes on under such conditions: 'A forest growing downwind of a stream of air laden with pestilential miasmas sometimes protects the terrain that is sheltered by it, whereas the terrain that remains open to the wind is exposed to illnesses of all sorts. The trees thus filter the infected air and purify it by removing the miasmas.' Moreover, it had also been shown that a supersaturated solution of sodium sulphate rapidly crystallized if the solution was exposed to air, but not if the air was first filtered through a layer of cotton wool. Finally, it was known to clinicians that barriers of cotton wool reduced the frequency with which wounds and other superficial injuries became infected. The decision by Schröder and von Dusch to try cotton wool as an air filter was thus not a bolt out of the blue, or the belated recognition of the obvious. It was the logical next step.

The apparatus, shown in Fig. 8.1, is basically the same as Schwann's, but with the incoming air filtered through a cotton wool plug instead of being heated. The airflow is arranged by means of a gasometer set up to act as an aspirator. Again, the organic solution, in this case an infusion of meat, is contained in a

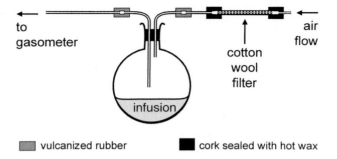

Fig. 8.1 Reconstruction of the apparatus used by Schröder and von Dusch. Air can enter the flask only via a barrier of cotton wool (see text)

flask sealed with a cork through which pass two tubes, one leading to the gasometer, the other to the cotton wool plug. Junctions in the various lengths of glass tubing are sealed with vulcanized rubber or with corks. All corks are waxed. The cotton wool is packed into a tube which communicates with the outside air through one of narrower gauge. Before the experiment is begun, the airtightness of the equipment is tested and any leaks sealed with hot wax. The infusion is then boiled until the rising steam has thoroughly heated all tubes including that containing the cotton wool. The gasometer tap is then set to establish a very slow passage of air through the cotton wool filter into the flask. This retarded airflow can be maintained continuously over many days by filling the gasometer with water morning and evening.

The first experiment was begun in winter (9 February 1853). A flask not shielded by a cotton wool plug served as a control. After 26 days the flasks were opened. The infusion exposed to the surrounding air had putrefied and stank; but there was no putrefaction in the infusion exposed only to air filtered through the cotton wool. The experiment was elaborated in warmer weather (April). In this case, there were three flasks. One was simply left open to the air. A second was sealed with a waxed cork perforated by a long glass tube that was left open at both ends so that air could enter the flask but only at a greatly reduced rate. The infusion in the third flask was boiled and, while it was still hot, a cotton wool stopper was stuffed into the mouth of the flask and covered with an additional thick layer of cotton wool. The infusion in the first flask had putrefied within two weeks and stank. In the second flask the surface of the infusion was covered with mould within 9 days, but when, 10 days later, the flask was opened, there was no smell of putrefaction, merely the smell of the mould. The third flask was opened after 24 days and showed neither putrefaction of the infusion nor any growth of mould. The results were therefore clear-cut: putrefaction did not occur if the air entering the flask was first filtered through cotton wool.

The importance of this finding has often been underestimated. First, it answered once and for all the objection that had been made to all experiments that had so far been conducted with heated air: that heating the air in some way spoilt it. It stretched credibility to suppose that the air could be spoilt by passage through the interstices of cotton wool. And secondly, the successful use of cotton wool as an air filter laid the foundation of a methodology that has survived to the present day. Glass tubes plugged with cotton wool form an essential part of the armamentarium of every modern microbiologist.

Schröder and von Dusch then began a series of experiments to see whether the same results held with other types of infusion. Spallanzani had already shown that to sterilize some infusions a temperature greater than that of boiling water might be required. But there had as yet been no systematic examination of this variability. This Schröder and von Dusch now undertook.

The experiment with filtered air was repeated with an infusion of malt containing a few hops from the local vinegar factory. After 8 days, controls without the cotton wool filters had putrefied and developed moulds. But where the air had been filtered, the infusion showed no putrefaction or growth of moulds after 23 days. Chemical analysis of the infusion at the completion of the experiment established that it had generated no alcohol.

When meat was tested without added water, putrefaction occurred even in the presence of filtered air and despite the fact that the meat had been heated to the temperature of boiling water. The explanation given for this anomalous result was that, without water, adequate heat had not reached the interior of the meat. This plausible explanation was strengthened by the results obtained when water was added to the meat. In an open flask, the meat infusion was putrid within 4 days and contained many infusoria, some of which resembled yeasts. In the infusion exposed only to filtered air, no putrefaction occurred even after 28 days. When the experiment was done with milk, however, the milk clotted, turned sour, and began to smell of cheese whether the air to which it was exposed had been filtered or not; but no moulds grew in the flask that had been supplied only with filtered air. No explanation of this anomalous result was offered.

Schröder and von Dusch drew the following conclusions: first, that the results obtained with a watery extract of meat were clear and consistent; second, that there were different kinds of putrefaction, some perhaps requiring air, others not, and some inhibited if the air was filtered, others not; and finally, that further experiments were needed to determine the sources of this variability. The authors also advocated that other materials, such as charcoal, lead sulphide, pumice, ground glass, and plaster of Paris should be tested for their ability to filter the air.

The results of this paper have often been misinterpreted. Due to the fact that putrefaction and growth of microorganisms occurred in some experiments in which the air had been filtered, the impression seems to have gained ground that the results obtained were ambiguous, if not unreliable. With hindsight it is easy to see that this is not the case. While the paper makes it clear that filtering the air is an effective means of preventing putrefaction, it also shows that with some organic substances this is not enough. The authors proposed to investigate these apparent exceptions further. To a modern eye this is a perfectly reasonable interpretation of the data and a laudable decision about what ought to be done next. In this paper, Schröder and von Dusch make no claims about spontaneous generation, but believers in that doctrine, when they noticed the work at all, were unimpressed by it.

At this point von Dusch left Mannheim, where the work had been done, to take up a position in Heidelberg, and there follow two papers by Schröder alone.[5,6] These are not widely quoted, but they are decisive. The first paper describes tests on a wider range of organic materials. In addition to milk and meat, experiments were done with egg white, egg yolk, blood, urine, starch, and some of their components. Except in the case of milk, egg yolk, and meat, the results were consistent: if, after boiling, the organic matter was exposed only to filtered air, no putrefaction occurred. In the case of milk, egg yolk, and some of the experiments with meat, putrefaction occurred despite filtration of the air. However, if the egg yolk was heated to a much higher temperature (160 °C in an oil bath) then putrefaction did *not* occur if the air was filtered. Urine, after heating, remained sterile for a year and a half in the presence of filtered air. Schröder concludes that almost all organic materials fail to putrefy if they are exposed only to filtered air, and moulds are never seen under these conditions. There is still no mention of spontaneous generation.

The final paper[7] in the series describes further experiments on milk, egg yolk, and meat, and introduces a new principle. Schröder had already shown that egg yolk failed to putrefy in filtered air if it was first heated to a much higher temperature than the boiling point of water. Now, however, he subjects his organic substances not only to increased heat, but also to increased atmospheric pressure. This innovation, like the introduction of the cotton wool air filters, gave rise, in due course, to another standard microbiological procedure. Sterilization of equipment and of contaminated material is to this day normally done in autoclaves, strong containers in which the objects to be sterilized are exposed to steam heat at high pressure.

Schröder heated the three anomalous substances, milk, egg yolk, and meat to 130 °C in an oil bath or exposed them in a pressure cooker to steam heat at a pressure of 3 atmospheres. After these procedures the refractory organic materials no longer putrefied in filtered air; and Schröder provides some preliminary information about the temperature and pressure required to produce this result. He makes one further important point. By means of a complicated piece of equipment that he has himself devised, he demonstrates that the putrefaction that takes place in milk when it is heated to no more than 100 °C is transferable to fresh milk even in the presence of filtered air.

The conclusion that Schröder draws from all this is that putrefaction and the generation of infusoria in experiments of this kind are due to the presence of seeds or germs introduced into the infusion by unfiltered air or present from the beginning but not inactivated by the temperature of boiling water. And now, at last, he raises the question of spontaneous generation. He declares himself an adherent of the view that all living things come from other living things (*omne vivum e vivo* again). He does not claim to have proved the non-existence of spontaneous generation, but he does claim to have demolished what experimental evidence there was in support of the idea. That claim is, in my view, entirely justified.

Notes

1. H. von Helmholz (1843), *Arch. Anat. Physiol. wiss. Med.*, 453.
2. H. Schröder and T. von Dusch (1854), *Ann. Chem. u. Pharm.* (new series) **13**: 232.
3. L. Pasteur (1861), *Annales des sciences naturelles* (partie zoologique) **16**: 4.
4. M. Becquerel (1853), *Compt. rend.* **36**: 10.
5. H. Schröder (1859), *Ann. Chem. u. Pharm.* (new series 33) **109**: 35.
6. H. Schröder (1861), *Ann. Chem. u. Pharm.* **117**: 273.
7. Ibid.

CHAPTER 9

The spontaneous generation of cells

In 1836 and 1837 Schwann presented an ingenious series of experiments that provided strong evidence against the view that microorganisms were the products of spontaneous generation. He was very circumspect in his conclusions mainly because of the criticisms to which his initial efforts had been subjected, but there is no doubt that he did not believe in the generation of life without divine intervention, or, at the very least, thought it unlikely. Yet in 1838 and 1839, a year or two later, he put forward a theory that in essence was no more than a special case of spontaneous generation.[1] He argued that the cells of the body were formed in the extracellular fluid by the aggregation of minute inanimate particles into larger units that eventually developed into whole cells. This was a remarkably abrupt volte-face. It is difficult to suppose that Schwann failed to appreciate that what he was proposing was indeed a form of spontaneous generation and that it was therefore diametrically opposed to the conclusions he had previously drawn. Schwann was a pious Catholic, and he must have found it satisfying to find that his earlier experiments supported the idea that life could not be created without supernatural intervention. But the mechanism that he was now proposing for the formation of cells was clearly envisaged as a natural process, not a supernatural one, and, whatever the details might turn out to be, it did involve the spontaneous transformation of the inanimate into the animate.

There is good evidence that Schwann was deeply disturbed by the apparent contradiction. Before publishing his famous monograph, he submitted it to the archbishop of Malines, primate of all Belgium, and it was only after the archbishop had approved of the text that Schwann proceeded with its publication. Whether the clerical imprimatur was granted because the archbishop judged Schwann's thesis to be plausible, or whether, in accordance with traditional church doctrine, permission to publish was granted because the thesis was held to be no more than a theory, is not known. In any case, there were some who saw at once that Schwann's model was inescapably an instance of spontaneous generation, complex though it might be. Robert Remak, of whom more later, claimed to have treated Schwann's ideas with profound scepticism from the moment they were published. 'As for myself,' he writes, 'the extracellular formation of animal cells struck me, from the very moment that this theory was propagated, as no less improbable than the *generatio aequivoca* of organisms.'

Schwann must at that time have been very suggestible, for he recounts that the origin of his theory was a conversation that he had with Matthias Schleiden (1804–81). Schleiden, at that time at the University of Jena, had just put forward a revolutionary theory of cell formation in plants.[2] He argued that the generative material was the viscous undifferentiated ground substance of which all cells were composed. Minute particles in this substance, he claimed, came together to form the nucleus, and when this and the material surrounding it were finally enclosed in a membrane, the generation of the cell was complete. This was the version of events, in its entirety a figment of his imagination, that Schleiden imparted to Schwann over dinner on the occasion of one of Schleiden's visits to Berlin.

It is amazing that Schwann, always meticulous, always circumspect, should have embraced Schleiden's model with such uncritical enthusiasm. Of course, the model had precedents. Milne-Edwards, who first proposed that the animal body was entirely composed of uniform globules, believed that these were produced by the growth of much smaller globules beyond the reach of the microscope. This view was shared by Dutrochet and, to some extent, by Raspail; but it was not universally accepted and was regarded by many as an unproven hypothesis.[3] According to Schwann, what he had been told by Schleiden at once brought to mind an observation that he himself had made on frog cells; and, full of excitement, he brought Schleiden back to his own small laboratory in Müller's department to show him the preparations he had made. From that moment on, Schwann devoted himself entirely to the problem of cell generation and, only a year after the publication of Schleiden's paper on plant cells, Schwann's famous monograph[4] on animal cells appeared.

However, as Remak notes, there was a fundamental difference between the models proposed by the two men. Schleiden envisaged that the formation of new cells from minute particles took place *within* existing cells, but Schwann, not finding any evidence of this in his material, proposed that the process took place in the extracellular fluid, by a process akin to crystallization. In the context of the debate about spontaneous generation, this distinction was crucial. For if new cells were formed within other cells, then the question of spontaneous generation did not arise. There were many who thought that 'daughter' cells were born within 'mother' cells, and Raspail, assuming that this was so, even sought to identify the intracellular organ that mediated the birth. What Schleiden was proposing was no more than a specific intracellular mechanism, but Schwann's variant raised an issue of much greater generality. For the extracellular fluid was inanimate, and Schwann was arguing that, by a completely natural process, it produced living cells. What there was about Schleiden's exposition that effected so rapid a transition in Schwann remains obscure.

Schwann's monograph was greeted with acclaim and widespread admiration. This was due, in part, to Müller's effective propaganda on its behalf, but perhaps mainly because, in an intellectual world still awash with *Naturphilosophie*, a theory that linked the animal and the plant kingdoms in one unifying principle was exactly what biologists wanted. The work had an arresting title: *Microscopic observations on the correspondence between animals and plants in their structure and growth*;[5] and it presented in compact form most of what was then known about the cellular composition of the tissues in both kingdoms. But above all, it was the unifying principle that was applauded, the principle that governed the formation of all cells whether animal or plant. It was this principle, and not mere structural similarities between animal and plant tissues, that Schwann himself regarded as the centrepiece of this theory; and it was this principle that endowed the inanimate extracellular fluid with the power to generate living cells.

Dissentient, or at least sceptical, voices began to be heard very soon after the publication of *Microscopic observations*, but they were nearly drowned out by the chorus of applause. This came not only from Schwann's colleagues or ex-colleagues in Müller's laboratory, but also from virtually all parts of Germany. In the literature of the early 1840s it is difficult to find any reference to the generation of cells that does not mention Schwann, or any mention of him that is not accompanied by high praise. Karl Rokitansky (1804–78), one of the most eminent figures in the development of modern pathology, adopted Schwann's views almost without modification. Rokitansky, who, for more than 30 years, was the head of the department of pathology at the general hospital in Vienna, published the first edition of his memorable work, the *Textbook of pathological anatomy*,[6] between 1842 and 1846, that is, in the wake of Schwann's monograph. Based on a massive number of autopsies, this *Textbook* laid the foundations for the present-day practice of hospital pathology, and it is remarkable to see how thoroughly the work is permeated by the model of cell generation proposed by Schwann. For Rokitansky, the origin of the disease states that he saw post-mortem lay in an anomalous admixture of the ingredients of the *blastema*, that primitive, structureless fluid from which, according to Schwann, all cells arose. There are few instances in the history of biology where an informed scientific community has succumbed so completely to a wave of fashion.

The first to put his doubts into writing was Franz Unger (1800–70). Unger was a native Austrian who worked first in Graz and eventually became the professor of botany in Vienna. In the course of a productive and influential life, he made more than one discovery of fundamental importance to the growth of botany as an experimental science. Unger first questioned Schleiden's model as early as 1841.[7] Even at that time, Unger thought that the commonest form of cell multiplication was that driven by cell division and not, as Schleiden proposed, by the *de novo* production of cells from a primitive ground substance. But Unger did not, in his initial paper on this subject, exclude other forms of cell generation.

It was known by 1841 that cell division could occur in multicellular as well as unicellular organisms. As mentioned earlier,

Abraham Trembley had discovered the phenomenon in micro-scopic polyps as early as 1744,[8] and Spallanzani had seen it in many species of unicellular infusoria. But it was not observed in multicellular organisms until 1832 when Barthélemy Dumortier (1797–1878) described it in silk-weeds.[9] A few years later, Hugo von Mohl (1805–72) gave a painstakingly detailed account of the process in the same material.[10] Neither Dumortier nor von Mohl believed that what they had discovered was the only mode of cell multiplication. Even after 1845 von Mohl still adhered to the view that cells were usually generated by the mechanism that Schleiden had proposed. By 1844, however, Unger had assembled enough evidence to challenge Schleiden directly. In a series of papers published in that year[11] he showed that growth of the plant stem was due to the accretion of cells arranged in rows and, by making measurements on the number and size of the cells formed at the growing tip, he reached the conclusion that new cells were formed by the division of existing cells into two. Unger's main objection to Schleiden's model was, however, that one did not find any intermediate forms, as one would expect if cells arose spon-taneously out of a primitive ground substance. But it was in the analysis of Schwann's variant of Schleiden's model that the issue of spontaneous generation of cells was to be settled, for it was Schwann who advocated that, in the animal body, the cells were formed by the aggregation of inanimate material in the extracel-lular fluid. The man primarily responsible for the demolition of Schwann's thesis was Robert Remak (1815–65).

Remak was a Polish Jew who was born in Posen and educated at the Polish Gymnasium there but spent his adult life in Ber-lin. His academic career was frustrated by his refusal to under-go baptism, for, at that time, no practising Jew had yet been appointed to a chair at the University of Berlin. In the conserva-tive circles that dominated the medical faculty, his advancement was not helped by the fact that he retained his loyalty to Polish nationalism and took an active part in various liberal organiza-tions both within the university and without. Despite his gener-ally acknowledged scientific achievements, he was never given a full-time academic appointment and, in mid-life, was obliged, for financial reasons, to take up private medical practice. He died, a deeply embittered man, at the age of 50.

As early as 1841, that is, only two years after the publication of Schwann's monograph, Remak, studying the generation of chick embryonic red cells, noticed that they multiplied by cell division, each mature cell dividing into two.[12] At no time did he see the cells arising *de novo* out of the extracellular fluid, nor did he see naked nuclei or intermediates that might be expected had red cells been formed in the manner that Schwann had suggested. Remak repeated this observation many times and demonstrated it annually to his students. By 1845, he had noticed the same form of cell division in embryonic muscle;[13] and by 1852 he had seen it in many of the organs of the developing embryo, and had come to the conclusion that it was the only form of cell multiplication to be found in the animal body.[14] In his great book of 1855, *Investigations on the development of vertebrates*,[15] Remak sets out the evidence and rejects Schwann's model in its entirety. The analogy that Schwann drew between the formation of the cell and crystallization Remak dismisses out of hand, demonstrating that in reality the two processes have nothing in common. More than anything else in Schwann's monograph, this analogy makes it clear that cell generation, in Schwann's mind, was a manifestation of spontaneous generation.

Remak's work was well known, but not generally accepted. No objections were raised to his experimental observations, but the great majority of his fellow-scientists were reluctant to believe that division of the mature cell into two was the only form of cell multiplication. This was true even of his former colleague Rudolf Virchow (1821–1902), but it was Virchow who, in the end, converted a sceptical world to Remak's point of view although Virchow never acknowledged that the point of view he was advocating was originally Remak's.

Virchow received his medical education at the Pépinière, Berlin's military medical school, and his training in pathology at the Charité, the city's main public hospital. After a highly successful period as professor of pathology at Würzburg, he returned to Berlin as professor of pathological anatomy and in that post soon acquired an unrivalled reputation both as an investigator and as a teacher. He eventually became one of the acknowledged leaders of European medicine and made contributions not only to pathology, but also to anthropology, public health, and even liberal politics. His place in popular medical history does not, however, rest so much on the originality of his contributions as on the association of his name with the doctrine that all cells are the progeny of other cells and are produced by the process of cell division. As late as 1854, Virchow was not altogether sure of this doctrine. Reviewing one of Remak's papers that year, Virchow could not conceal his doubts. A year later he had become not only a convert to the uncompromising position adopted by Remak, but also an enthusiastic advocate of it. What brought about a conversion so rapid and so complete unfortunately remains obscure.

In 1855, the year in which Remak published his book on vertebrate development, Virchow wrote an editorial in the journal that he himself had founded. It bore the title *Cellular-Pathologie*[16] and argued vehemently for pathology as an experimental science based on the behaviour of cells; the essential principle that was to guide this remade study of disease was that new cells were generated by the division of existing cells into two, and in no other way. This same principle is the leitmotiv of Virchow's celebrated book the *Cellularpathologie*,[17] which appeared in 1858. Virchow, like Remak before him, saw Schwann's view of cell formation as an instance of *generatio aequivoca*, an idea that he treats with contempt. The *Cellularpathologie* was an immediate success, ran through three editions in four years, and was translated into all the major European languages. The memorable phrase that Virchow adopted as his motto, *omnis cellula e cellula* (all cells come from other cells), carried all before it and is still quoted in every recital of Virchow's place in the history of medicine. By the time Virchow's demolition of Schwann had percolated into medical establishments throughout Europe, no one argued any longer that the cells of the body could be formed by spontaneous generation. There remained only one group of organisms for which spontaneous generation remained a plausible option—the infusoria, and especially the smallest of the animalcules that Leeuwenhoek had described.

Notes

1. See Chapter 6, note 10.
2. M. J. Schleiden (1838), *Arch. Anat. Physiol. wiss. Med.*, 137.

3. See Chapter 2, note 4.

4. See Chapter 6, note 10.

5. See Chapter 6, note 10.

6. K. Rokitansky (1842–6), *Handbuch der pathologischen Anatomie.* Vienna .

7. F. Unger (1841), *Linnaea.*, 385.

8. See Chapter 4, note 4.

9. B. C. Dumortier (1832), *Nova Acta Phys.-Med. Acad. Caesar. Leopold-Carolinae Nat. Curios.* (Part 1) **16**: 217.

10. H. von Mohl (1837), *Allgemeine bot. Zeitschr.* **1**: 17.

11. F. Unger (1844), *Bot. Zeit.* **2**: cols 489, 506, 521.

12. R. Remak (1841), *Z. Ver. Heilk. Pr.* **10**: 127; *Canstatts Jahresber. ges. Med.* **1**: 17.

13. R. Remak (1845), *Frorieps Neue Notizen* **35**: 305.

14. R. Remak (1852), *Arch. Anat. Physiol. wiss. Med.*, 47.

15. R. Remak (1855), *Untersuchungen über die Entwickelung der Wirbelthiere.* Berlin.

16. R. Virchow (1855), *Arch. path. Anat. Physiol. klin. Med.*, **8**: 3.

17. R. Virchow (1858), *Cellularpathologie.* Berlin.

CHAPTER 10

A disagreement in the Académie

B y the end of the 1850s there was virtual unanimity in the Académie des sciences in Paris that spontaneous generation did not exist. Milne-Edwards, in a report communicated to the Académie in 1859,[1] began by saying that there were then so few zoologists who believed in spontaneous generation that he would have been afraid of wasting the time of the Académie if it weren't for a new communication on this subject by M. Pouchet. Félix Archimède Pouchet (1800–72), director of the Natural History Museum at Rouen and corresponding member of the Académie, had announced that he had observed the spontaneous generation of animal and plant microorganisms under strictly controlled experimental conditions.[2] The ensuing debate, between Pouchet and his collaborators on the one hand, and Louis Pasteur (1822–95) (Fig. 10.1) and several other distinguished members of the Académie on the other, has been a set piece for historians of science.[3] But they have been concerned mainly with the religious and social background of the debate; with the presentational skills of the participants; and with the impartiality, or lack of it, of the Académiciens. Little attention has been paid to the probative value of the evidence presented.

Pouchet was a learned and religious man. Before he became involved in the question of spontaneous generation, he had written a substantial work on Albertus Magnus and the state of science in the Middle Ages; and he always maintained that his belief in spontaneous generation in no way conflicted with his Catholicism. Pasteur, who was also a pious and conservative Catholic, similarly concluded that his dismissal of spontaneous generation was entirely compatible with Catholic doctrine. It is

Fig. 10.1 Louis Pasteur (1822–95)

therefore unlikely that a preconceived religious assumption prompted Pouchet to initiate his experiments on the creation of life. We do not know what the stimulus was. His first publication on the subject was the report that he made to the Académie in 1858.[4] This *compte rendu* bore the startling title 'Note on the plant and animal proto-organisms born spontaneously in artificial air and oxygen'. It created a stir not only in the ranks of the Académie, but also in the learned world of Paris at large. There was an immediate response from five of the most distinguished members of the Académie, and the reports of what they had to say appeared in the same journal afterwards.[5] These reports are of special historical interest because they allow us to see Pouchet's experiments not only with the acuity of hindsight, but also as other scientists saw them at the time.

Pouchet opens his paper with a claim that would have raised eyebrows in the Académie, just as it does today. He asserts that he has carried out all the experiments that had so far been done on the subject of spontaneous generation up to and including those of Schulze and Schwann. Given the vast amount of work that Spallanzani and others had done on the subject, this assertion can only be regarded as an idle boast. In the case of Schulze and Schwann, Pouchet says that he repeated their experiments *exactly*, but in the same breath adds that he 'even modified them' to ensure greater precision; in what way is not stated. The main burden of Pouchet's paper is that he saw plant and animal microorganisms growing in boiled organic medium even when the air supplied to it had passed through sulphuric acid, or through a labyrinth of porcelain fragments, or through asbestos heated to redness. Schwann had himself admitted that he had obtained a similar result, but only occasionally and under rather special circumstances. But the conclusion that Pouchet drew was diametrically opposed to Schwann's. Schwann regarded growth of microorganisms under these conditions as exceptional; Pouchet regarded it as the rule.

Since it was, by 1858, commonly held that organisms were brought into the vessel containing the boiled organic solution by the flow of air, Pouchet decided to replace atmospheric air with an artificial mixture of oxygen and nitrogen produced chemically (21 parts oxygen and 79 parts nitrogen). He also examined the effect of pure oxygen. The gases were prepared by M. Houzeau, a professor of chemistry. The experimental procedure is described in some detail. A one-litre flask is filled with boiling water, hermetically sealed, and then inverted into a cuvette containing mercury. When the flask has cooled completely it is opened under mercury and half a litre of pure oxygen introduced. Immediately afterwards a little truss of hay (10 grams) is inserted into the flask, again under mercury. The hay is taken from a closed flask which has been heated to 100 °C for 30 minutes in an oven. The flask containing the artificial air and hay is sealed with a ground glass stopper and then taken out of the mercury. The neck of the flask and the stopper are finally covered with a layer of varnish and the highly toxic vermilion (mercuric chloride). Eight days later Pouchet sees clumps of fungus growing in the flask. They seem to him to be a species of *Aspergillus*, and this is later confirmed by M. Montagne, an expert mycologist, who decides that the fungus is actually a new species and names it *Aspergillus Pouchetii*.

In the last part of his communication, Pouchet discusses the possibility that the 'germs' of fungi might not be killed at 100 °C. This he tests on the germs of *Penicillium glaucum*. These are normally 'perfectly spherical', but after 15 minutes in a tube heated with a spirit lamp, they are deformed, have almost doubled their volume, and, in some cases, have lost their contents. *Aspergillus* is found to be even more susceptible to heat than the *Penicillium*.

Pouchet's paper was followed by a second, this time by Pouchet and Houzeau.[6] The procedure is essentially the same as in the previous experiments, but a five-litre flask with a ground glass stopper is used. The mixture of oxygen and nitrogen is introduced into the flask which is finally sealed with copal (a resin) thickened with vermilion. This time growths of both *Aspergillus* and *Penicillium* are seen and numerous animal infusoria of which seven could be identified by morphological criteria. Experiments done 'by one of us' (which one is not specified, but it must have been Pouchet) showed that the germs of the infusoria were also destroyed by a temperature of 100 °C.

Pouchet's experiments and the conclusion that he drew from them were rejected by the members of the Académie best qualified to judge them: Milne-Edwards, Payen, Quatrefages, Claude Bernard, and Dumas. Henri Milne-Edwards, whose early attempt to identify animal cells has already been mentioned, was now professor of zoology at the Muséum national d'histoire naturelle in Paris. He was also the editor of the zoological section of the *Comptes rendus*. Anselme Payen, noted for the hugely important discovery that cellulose was the essential component of plant cell walls, spoke for the botanists. Armand de Quatrefages, professor of anatomy and professor of anthropology at the University of

Paris, had, some years previously, made a special study of early embryonic development in limnetic animals. Claude Bernard, professor of experimental physiology at the Collège de France, was one of the most eminent physiologists of the nineteenth century and a founding father of modern physiological research. And Jean-Baptiste Dumas was a distinguished chemist who eventually became the *Secrétaire perpétuel* of the Académie des sciences. I have given these skeletal biographical details to show that the men who gave voice to their objections were not a coterie of closed minds as they have sometimes been depicted. They were highly respected scientists who, collectively, represented zoology, botany, embryology, physiology, and chemistry; and, as becomes apparent in their criticisms of Pouchet, they all had more than adequate qualifications to assess the power of his experiments on spontaneous generation. Indeed, it is doubtful whether the membership of the Académie could have provided representatives with better qualifications.

In the introduction to his communication, Milne-Edwards points out that after the work of Redi, spontaneous generation had few supporters until Leeuwenhoek's discovery opened up the new world of microscopic animalcules. It was the obscure origin of these minute creatures that resurrected the theory. However, Milne-Edwards stresses that if one wants to explore spontaneous generation by means of experiments of the kind done by Redi (open and closed vessels), then, with infusoria, one needs to take special precautions. This advice must have struck the biologists in the Académie as a platitude, and Milne-Edwards would not have given it unless he was far from satisfied with the precautions taken by Pouchet. Given that Pouchet does not describe how he ensured the sterility of his artificial air or how he maintained the sterility of his truss of hay, this is not surprising. But Milne-Edwards's main objection is that the temperature to which Pouchet had heated his truss of hay in the first place would not have rendered it free from living microorganisms or their germs. To begin with, the hay would have been surrounded by static air, which is a poor conductor of heat, so that the external temperature might not have been reached within the truss itself. Even if a heat equilibrium had been reached, there is still a problem if the microorganisms are dry. For although it had been shown that animals die if the ambient temperature reaches the point at which albumin coagulates, it had also been shown that some microorganisms survive higher temperatures if they are first desiccated. For example, when dry, tardigrades (minute animals that inhabit mosses) survive several hours at much higher temperatures than those used by Pouchet. Milne-Edwards had seen animalcules survive prolonged heating at a temperature of 120 °C. Like many others, he had himself carried out experiments with boiled organic infusions in open flasks and in flasks hermetically sealed by means of a flame. He found that animalcules grew in the open flasks, but not in those that had been hermetically sealed. In conclusion he explains that his comments were made so that young physiologists should not be misled.

In his contribution to the debate, Payen draws attention to the discovery made by himself and de Mirbel that deterioration of bread may be brought about by infestation with a primitive plant. When the spores of this plant were examined, it was found that they survived a temperature of 120 °C, but became discoloured and failed to germinate if the temperature was raised to 140 °C.

The communication by Quatrefages begins with the statement that it is generally admitted that spontaneous generation does not exist. Those who still support the theory often claim that atmospheric air could not possibly contain infusoria or their germs in a high enough concentration to account for the results that had been obtained. But Quatrefages draws attention to the findings of Boussingault, who examined what was left on a filter that had been exposed to a downpour of rain. An assortment of easily recognizable animal and plant forms was detected, and Quatrefages considers that very exacting procedures would be needed to exclude these from one's experiments. He then digresses to discuss the origin of intestinal worms. These, he emphasizes, are not produced by spontaneous generation but by sexual conjugation, although the life-cycles of these animals may contain phases of asexual reproduction. This discovery he attributes to Küchenmeister and van Beneden. In conclusion, he refers to Ehrenberg's observation that infusoria multiply very rapidly, so that such enormous numbers do not need to be present in atmospheric air to explain the extent of the growth seen in organic infusions.

Claude Bernard's report is entirely experimental. He relates that following many earlier studies that he had made on the role of sugars in the growth of microscopic plants, he carried out the following experiment begun in September in the year preceding that of the first publication of Pouchet's work. Fifty millilitres of a dilute solution of gelatin in water are introduced into two flasks and a trace of cane sugar added. The necks of the flasks are drawn out to facilitate sealing at a later stage. One flask is fitted with a porcelain tube into which fragments of porcelain are inserted. The flasks are boiled for a quarter of an hour, and, as the boiling stops and the steam condenses, air is slowly drawn into them. In one case, the air enters only via the porcelain tube, which is heated to redness; in the other, it enters without passing through a porcelain tube. The flasks, on cooling, are sealed hermetically by annealing the narrowed region of their necks. After 10–12 days, the flask into which the air entered without passing through the heated porcelain tube shows moulds on the surface of the fluid. The moulds increase in size for about six months and thereafter remain stationary. The flask that received only heated air shows no growth of any kind. When the flasks are finally opened, the fluid in the one that received unheated air is clearly putrid and stinks; the fluid in the other shows no sign of putrescence. The mould was later identified as *Penicillium glaucum*. Bernard's closing remark was either commendably restrained or tongue in cheek: 'It can be seen that this experiment, like those mentioned previously, does not favour the theory of spontaneous generation.'

Dumas recalls that some thirty years previously he was stimulated to examine the question of spontaneous generation because of the work of Fray, who had reached the same conclusion as Pouchet. Dumas had found that when organic materials in artificial air or artificial water were enclosed within glass tubes and heated to 120–130 °C, they did not subsequently generate either plant or animal microorganisms. But both were produced when atmospheric air was allowed to enter the tubes. Dumas reiterates the point made by Milne-Edwards: in the dry state some tardigrades resist 140 °C, and some spores resist 100 °C even when wet. In Dumas's view, Pouchet's experiments do not, therefore, establish the existence of spontaneous generation, for one would expect to see growth of microorganisms from time to time when the temperature is raised merely to the boiling point of water.

It is thus clear that, as a whole, the members of the Académie did not believe in spontaneous generation, and they were certainly not persuaded to change their minds by the evidence produced by Pouchet. It has been argued[7] that the Académiciens who criticized Pouchet were biased, but this, it seems to me, is a highly one-sided interpretation of the historical situation. Pouchet's critics had not reached their conclusions about spontaneous generation as a result of prejudice, religious, social, or any other; their attitude was based on an assessment of two hundred years of meticulous experiment. The objections that they raised to Pouchet's work were valid at the time and remain valid to the present day.

Pouchet's reply to these criticisms appeared in the *Comptes rendus* in the following year.[8] It was accompanied by an editorial footnote which explained that the Académie, not wishing to curtail the response of M. Pouchet to the criticisms made by several of its members, had decided to print his communication in full, although it exceeded the normal limit permitted in the *Comptes rendus*. This footnote makes it clear that the Académie, despite the scepticism of most of its members, was anxious that Pouchet should be given a fair hearing. His reply, impossibly obsequious by modern standards, must have seemed unusually courteous even in the polite and formal atmosphere of Académie meetings. He flatters even those Académiciens who opposed him.

After an elaborate exordium, Pouchet deals with each of his critics in turn. In answer to Milne-Edwards he makes three points. First, the stalks of the hay used were very fine, so that the truss was 'undoubtedly' permeated by the temperature applied. This is said to have been checked by M. Houzeau with a thermometer, although what kind of thermometer it was that penetrated the interstices of the hay or how this was done is not stated. What is more, Pouchet has since repeated the experiment with hay heated to 200–250 °C and even with hay reduced to ash, but animalcules still appear. Pouchet's second point is a defence of his technique. This takes the form of a series of *ad hoc* arguments rather than additional evidence. Whenever, he argues, growth of microorganisms is observed in a hermetically sealed vessel, the objection is raised that atmospheric air has somehow leaked into it. But even if this happened, the atmospheric air would not produce the population of organisms that one sees in the vessel. Higher order infusoria, such as one sees in air, are not seen in hermetically sealed vessels, and, on the other hand, one sees microorganisms there that are not seen in air. This argument would hardly have made much impression on his audience, for they were well aware that what one sees in air depends on the size of the sample, and what one sees in the organic solution reflects the relative growth rates of the microorganisms involved, some of which, as Quatrefages mentioned, grow very rapidly.

Pouchet's third point is a direct criticism of Milne-Edwards's own experiment. If Milne-Edwards sealed his flask while the infusion was still boiling, as Pouchet understands he did, then he would have created a partial vacuum which would not support life. Moreover, if Milne-Edwards believes that 100 °C does not kill all the germs in the organic solution, then how does it happen that his flasks are not full of animalcules? These are debating points that throw little light on the real situation. Numerous experimenters had sealed their flasks after the solution had cooled and still failed to detect any growth of microorganisms if the external air was heated before it entered the flask. And Pouchet's critics knew full well that growth of microorganisms in boiled solutions occurred only sporadically, a phenomenon that they believed was attributable to the occasional presence in the solution of microorganisms that did indeed resist 100 °C. Pouchet's reply to Milne-Edwards ends with a prestige argument. He lists a number of distinguished figures who, in the past, had supported spontaneous generation. But all these names were well known to the biologists in the Académie, and they had already assessed and rejected the arguments that these eminent men of an earlier generation had adduced.

In reply to Quatrefages, Pouchet has this to say. Quatrefages had mentioned that he had observed what appeared to be dust particles in the air, but when these particles were immersed in water, they were seen to be the eggs of animalcules. Pouchet answers that distilled, boiled, or filtered water left open to the air in large cuvettes does not become infested with infusoria, but if a little organic material is added, the water soon teems with animalcules. Admittedly, it could be argued that the development of the egg might require nutriment, but Pouchet claims that nutriment is not needed for the earliest stages of animal development. The mind boggles, he claims, at the density and variety of organisms that must be present in air to produce the observed results. An eminent, but unnamed, zoologist had apparently calculated that a drop of water must contain about five hundred million such organisms. Pouchet's opponents would have known at once that these calculations were worthless, for the growth rates of the organisms involved, as yet undetermined, could not have been taken into account. Pouchet agrees with Quatrefages that large infusoria might undergo sexual reproduction, but he regards this as an infrequent occurrence, and he completely rejects the idea that microorganisms might multiply by dividing into two. He concedes that this phenomenon has been observed, but he regards it as a developmental abnormality analogous to the formation of double foetuses in mammals and birds. Again, he cites eminent figures who, in earlier times, had supported his views.

In Pouchet's opinion only the experiments of Schulze and Schwann offer serious evidence against spontaneous generation. (Pouchet consistently misspells Schulze, as does Pasteur; few of the Académiciens seem able to spell Leeuwenhoek correctly.) But Pouchet contends that even these experiments were carried out carelessly ('avec fort peu de précision'), a slur that could hardly be further from the truth. In support of his own position, he quotes the remarks of Bérard, whose reputation he considers comparable to that of Claude Bernard ('possédant aussi une illustre renommé'). Bérard accepts that spontaneous generation does exist and concludes that even if Schulze and Schwann were right, that would only prove that animalcules cannot grow in air spoilt ('tourmenté') by sulphuric acid or red heat. For good measure he adds that if it were true that there were microorganisms that survived $100\,°C$, then the experiments of Schulze and Schwann would be utterly meaningless ('nul'). Pouchet does not discuss Claude Bernard's experiments in any detail, for the 'spoilt air' argument he had used against Schulze and Schwann applied equally well to Bernard. But Pouchet does go on to say that the putrid smell of the infusions that had been exposed to atmospheric air, and the absence of any such smell in the infusions exposed to heated air, confirmed that the heated air had been spoilt. This remark is unlikely to have enhanced Pouchet's reputation among the Académiciens.

Payen's criticism that some spores easily resist the temperature of boiling water Pouchet answers by citing Morren's observation that 45 °C is hot enough to kill infusoria. What relevance this has to the heat-resistance of Payen's spores is not clear. And Dumas, who also stressed the heat-resistance of some microorganisms and their spores, Pouchet answers by referring him to *Hétérogénie*,[9] a book about spontaneous generation that Pouchet published in that same year. In *Hétérogénie*, he asserts, there are experiments with putrescible materials that were heated to 220 °C, and in artificial water.

I have given a rather elaborate account of Pouchet's original paper, the criticisms made by several distinguished Académiciens, and Pouchet's reply to these criticisms. This, it seems to me, must be done if an informed assessment is to be made of the strength of his case and of the Académie's reaction to it. Some modern historians have taken the view that in some respects Pouchet's arguments had merit and that he was unfairly treated by the Académie.[10] I do not believe that either proposition can be sustained. Pouchet's position was, from the beginning, an idiosyncratic one, and powerful experiments would have been necessary to shake the almost unanimous view of the Académie that spontaneous generation did not exist. Pouchet's experiments were not powerful. They exhibited errors of technique with which the biologists in the Académie were thoroughly familiar; and on the basis of these experiments Pouchet made a heroic claim that the Académiciens could not possibly accept. His reply to their criticisms was evasive, excessively flattering to those who in the past had supported his point of view, and gratuitously disparaging about the experiments of those who had reached the opposite conclusion. And the tone of the reply is so obsequious that it can only have repelled the Académiciens whose approbation Pouchet was seeking. Given the attitude that prevailed in the Académie at the time and the very poor impression that Pouchet's first paper obviously created, it says a great deal for the propriety of the editors of the *Comptes rendus* that they made extra space available to ensure that Pouchet's reply could be printed in full. It has, nonetheless, been contended that the Académie behaved unjustly at a later stage. Whether this was so or not is a matter that will be considered presently.

It was after the publication of Pouchet's reply to his critics that Pasteur made his first contribution to the controversy. This appeared in the *Comptes rendus* in February 1860 and was followed by four further reports in that journal, published later in the year and in the early part of 1861.[11] The contents of these *comptes rendus* were expanded in a lengthy paper (over 90 pages) which appeared in the *Annales des sciences naturelles* in 1861;[12] and a popular exposition of the work was given at a *Soirée scientifique de la Sorbonne* in 1864.[13] It is important that the purposes of these three avenues of communication be distinguished. The reports in the *Comptes rendus* are factual accounts of scientific work completed or in progress. The paper in the *Annales* is the published version of an essay that Pasteur submitted to the Académie in an open competition for the Alhumbert prize. And the exposé at the Sorbonne was a public lecture delivered to a mixed audience that included non-scientists as well as scientists. These publications of Pasteur's have attracted a good deal of criticism in recent years.[14] It has been argued that his presentation of the data is one-sided, that his style is excessively rhetorical, indeed grandiloquent, and even that he often 'failed to conform to ordinary notions of proper Scientific Method'.[15] I think these criticisms have arisen for two reasons: first, because of a failure to appreciate the respective functions of these very different publications; and second, because of an idealized, quite unreal, view of the Scientific Method.

There is nothing grandiloquent about the reports in the *Comptes rendus*. The presentation is admittedly forceful, but the papers do not contain much more than a record of Pasteur's experimental findings and the conclusions that he draws from them. The essay in the *Annales* is something else altogether. Being an entry for a prize, it is a piece of advocacy as well as a record of experimental findings. The review of the relevant literature with which the essay begins is clearly tendentious and, on occasion, inaccurate. And in each of the nine chapters into which the essay is divided, Pasteur seeks to make a case. He disparages the experiments of those who claimed to have observed spontaneous generation, and he often belittles the work of predecessors who, like himself, had failed to find any evidence of spontaneous generation. He does not report any results of his own that might cast doubt on his general conclusions, although it is difficult to believe that, in a long series of investigations, he never observed any contradictory results. But, in a prize essay, designed to convince his judges of a particular point of view, it would have been surprising had aberrant results been reported. The circumstances that mitigate Pasteur's lack of objectivity in the *Annales* paper apply even more cogently to the lecture at the Soirée. This is, and was meant to be, a histrionic performance designed to be understood, and enjoyed, by an educated lay audience; impartial analysis of conflicting experimental data would have been completely out of place. But whatever reservations one might have about Pasteur's one-sided presentation in the *Annales* paper, they should not be permitted to obscure the power of the experiments themselves. In most cases, these experiments were decisive. No one can read the *Annales* paper without admiring Pasteur's ingenuity and his experimental skill.

The first chapter in the essay is the historical introduction; this is not intended to be serious historiography and does not require serious historiographical analysis. It is worth noting nonetheless that Pasteur does quote the paper by Schröder and von Dusch and the first of the two papers by Schröder, but not the second, which is the one that contains the definitive information about the high temperature and high pressure required to sterilize some organic materials. It seems probable, given the fact that both Schröder's second paper and Pasteur's *Annales* paper were published in the same year, that Pasteur did not yet know of Schröder's second contribution. However, it cannot be overlooked that in his own later publications on this theme Pasteur still does not give Schröder the credit that he undoubtedly deserves. Pasteur states that he was led to the study of spontaneous generation by his own work on fermentation, for fermentation always appeared to be associated with the presence of microorganisms. Dumas advised him not to touch the problem.

In Chapter 2, Pasteur tackles the most important of Pouchet's claims, namely that atmospheric air simply does not have enough organisms in it to account for the results obtained with organic solutions. In Pouchet's view, most of the microscopic bodies that one sees in air are inorganic or organic particles, not seeds or germs. Pasteur's answer to this is to collect the particles in a known volume of air, estimate how many there are, and, by microscopic examination, determine their nature. This he does by passing a tube, plugged with a soluble cotton wool barrier, through the frame of a window opening onto the rue d'Ulm or through one opening onto the garden of the Ecole normale. The tube is connected to a three-way tap set so that the flow of water draws the outside air through the cotton wool barrier. The volume of air passing through the cotton wool in a given time can be calculated from the amount of water that flows through the system. The cotton wool plug is then removed from the tube and dissolved in a mixture of alcohol and ether. (In the experiments reported in the *Comptes rendus* paper, concentrated sulphuric acid was used. This was abandoned presumably because the strong acid destroyed some of the air-borne particles, or because cotton wool soluble in alcohol and ether was not available.) The particulate material released from the cotton wool plug is allowed to settle for 12–20 hours and the supernatant is decanted. The particles are then transferred to a watch-glass, evaporated to dryness, and stained for microscopical examination.

Pasteur gives a detailed description of the organisms that he sees, and illustrates them. They include slime-moulds, primitive fungi, and yeasts. He calculates that several thousand 'organized' particles are collected on the cotton wool plug in 24 hours when the air is drawn from the street at a rate of about one litre per minute. These experiments are completely convincing. There can have been little residual doubt in the minds of interested Académiciens that Pouchet was hopelessly wrong in his estimate of the number of microorganisms in atmospheric air. But, as we shall see later, Pouchet himself was not convinced.

Chapter 3 describes experiments in which the atmospheric air is scorched ('calciné') before being permitted to enter the flask containing the organic solution. In this case, the solution is simply water to which sugar and an extract of brewer's yeast are added. A small platinum tube is intercalated into the line of rubber tubing that provides the only avenue of communication between the flask and the outside air. The platinum tube is heated to red heat and the organic solution boiled for two to three minutes. As the solution cools, air enters the flask but only via the red-hot platinum tube. The flask was then sealed. No flasks treated in this way subsequently showed any sign of putrefaction.

Pasteur then inveighs against the use of mercury in experiments of this sort. He points out that the surface of mercury in a cuvette always collects dust particles, so that if one tries to insert something into an inverted flask through a barrier of mercury, one inevitably inserts a layer of dust and air with it. This is precisely what Pouchet had done with his truss of hay, so if Pasteur's criticism is valid, Pouchet's experimental technique is irrevocably undermined. (Pasteur labours this point in his *Soirée* lecture and gives a dramatic visual demonstration of the unavoidable layer of dust associated with mercury.) Pasteur does not specifically refer to Schwann's use of mercury, but he does cast doubt on all previous experiments in which a mercury barrier was used. Actually, this criticism does not apply to Schwann's work; in Schwann's experiments the cuvette of mercury did not serve as a protective barrier for objects that he wished to insert *into* the flask, but merely as a trap for air escaping *from* the flask. What is more, in order to destroy any microorganisms that might be present on the surface of the mercury, he covered it with a layer of oil and introduced some corrosive sublimate into the interface between the oil and the mercury. In arguing that all experiments involving the use of mercury were suspect, Pasteur may well have been overstating his case; but his criticism of Pouchet's use of mercury was undoubtedly justified. It should, however, be noted that none of the experiments that Pasteur describes in this chapter answer the perennial objection that scorching the air in some way spoils it. The definitive answer to that question was provided by Schröder.

In the next chapter, Pasteur provides decisive evidence that the particles collected from the air onto cotton wool can cause the putrefaction of organic solutions and the growth of microorganisms within them. In this case, the platinum tube, intercalated as before into the rubber tubing, contains the cotton wool plug on which the particles from the air had been deposited. Again, the platinum tube is heated to red heat, and the organic solution containing nutrients suitable for the growth of microorganisms is boiled. As the solution cools, atmospheric air, provided by the three-way tap, is allowed to enter the flask, but only through the cotton wool plug enclosed in the heated platinum tube. This procedure is repeated two or three times and the plug is then slid into the flask. The nutrient solution, in due course, undergoes putrefaction, again associated with the growth of various micro-organisms. Some of these are identified and illustrated. Repetition of this experiment always gives the same result. Pasteur's conclusion is uncompromising:

> In the light of these results, confirmed and extended by those described in the following chapters, I regard it as proven with mathematical certitude that when water containing sugar and albumin is brought to the boil and then exposed to ordinary air, the organisms that are formed within it originate in solid particles suspended in the air.

While the judges for the Alhumbert prize must have agreed with this conclusion, the experiments with the heated cotton wool plug do leave one or two questions unanswered. Pasteur asserts that all the air entering the flasks is heated well into the interstices of the cotton wool which, however, still retains its dust and hence its living organisms. But no attempt appears to have been made to measure the temperature in the interstices of the cotton wool. This omission does not cloud the conclusion that the solid particles in the air are responsible for the putrefaction, but, given the insulating power of cotton wool and the criticisms that Milne-Edwards had made of Pouchet's attempt to heat a truss of hay, it could be argued that some insufficiently heated air also enters the flask. This gloss would probably have been of importance only to those whose commitment to spontaneous generation was irreversible. For if it is true that microorganisms are generated in completely scorched air, then the scorched air cannot have been spoilt. But if some unscorched air, containing its 'force vitale', enters the flask together with the scorched air, then the spoilt air argument still survives.

The remaining chapters in the essay contain extensions of the basic findings that Pasteur regards as established beyond doubt, but they also introduce one technical advance that has since become almost legendary. Chapter 5 reports the same sort of experiments conducted with urine or milk instead of sugared water containing a little yeast extract. Again, Pasteur finds that urine boiled and then shielded from atmospheric air does not putrefy. But milk came as a surprise. Boiled milk shielded from the air nonetheless clotted and produced microorganisms. This was the first indication in Pasteur's essay that what one was boiling might be of importance, and that there were, despite all he had said, situations in which organisms grew in organic solutions despite boiling and exclusion of atmospheric air. Pasteur examined the organisms that grew in boiled milk and found that they were resistant to 100 °C. They could, however, be eliminated if the temperature of the milk was raised to 110 °C. This latter finding eliminated the anomaly posed by the behaviour of milk, but it also raised questions about all experiments in which the organic solution was merely boiled. It is perhaps surprising that Pasteur reports no anomalous results in any of his previous experiments. As mentioned previously, similar findings with urine and milk had already been reported by Schröder, who eventually resolved the problems associated with heat-resistant organisms by subjecting the organic solutions to much higher temperatures and pressures.

Chapter 6 introduces what is now generally known as the 'swan-necked' flask (Fig. 10.2). Pasteur devised a number of different flasks in which the necks were drawn out to provide

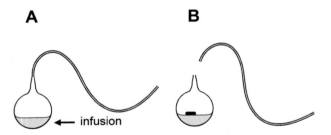

Fig. 10.2 Pasteur's drawing of the 'swan-necked' flask. In the intact flask the infusion remains sterile, but if the swan-neck is broken off the infusion putrefies (see text)

chambers or convolutions that might trap air-borne microorganisms. But it appears to have been the elegant simplicity of the swan-necked flask that gained it a place of such eminence in medical history. The idea behind all these flasks was to show that the flask could remain open to the air without putrefaction of its contents if the particles in the air were allowed to settle in the convolutions or other distortions of the elongated neck. Pasteur showed that a boiled organic solution contained in a swan-necked flask which was open to the outside air would remain free of putrefaction and microorganisms indefinitely if the flask was not shaken; but if the elongated neck was broken off, and a larger opening to the flask thus provided, putrefaction occurred. Pasteur's conclusion was that atmospheric air could be allowed to enter the flask without causing putrefaction if the particles in the air were first trapped.

Pasteur considered that the experiments with the swan-necked flask, coupled with the work that he had already described, delivered a mortal blow ('coup mortel') to the theory of spontaneous generation. But supporters of that theory might still have had their reservations. Schwann had introduced the use of the pump into experiments of this sort in order to provide the flask with an adequate and continuous supply of atmospheric air. This, it had been argued, was essential for the spontaneous generation of microorganisms. But if the air in the sinuous neck of the swan-necked flask permitted the air-borne particles to settle, then the flow of air through that narrow conduit must have been exceedingly slow, not much faster, indeed, than the rate of diffusion. It could therefore still be maintained that the flasks with elongated necks, although open to the outside air, did not provide an adequate supply of it. Nonetheless, the swan-necked flask continues to be regarded, even today, as the ultimate nail in the coffin of spontaneous generation.

Next, Pasteur describes experiments demonstrating that the microorganisms in the air were not homogeneously distributed. If flasks containing boiled organic solutions were opened briefly in different atmospheric conditions, the number that showed putrefaction varied. Putrefaction occurred in only a few of the flasks that were opened after a shower of rain, and in only one out of ten opened in the cellars of the Observatory, where the air was undisturbed. Similar experiments conducted at different altitudes in the Jura mountains showed that the higher one went, the less likely it was that the flasks would show putrefaction. Pasteur gives numbers for each of these experiments, but is aware of their statistical weakness, since only a few flasks were examined and the incidence of putrefaction varied. Appropriate statistical methods for this kind of experiment were not in fact developed until the twentieth century.

In his penultimate chapter, Pasteur merely extends his observations on the heat-resistance of spores and other air-borne microorganisms; some, he now finds, resist temperatures of up to 120 °C. In his final chapter, however, he makes a point of great theoretical importance. Ever since the time of Needham and Buffon, spontaneous generation was envisaged as a process taking place in organic solutions. Buffon had argued that the templates capable of generating living forms were formed by the aggregation of *organic* particles, and although later investigators suggested many variations on this general theme, no one envisaged that a reaction of this sort could take place in inorganic solutions. Pasteur now showed that, under appropriate conditions, both plant and animal microorganisms could indeed grow in inorganic solutions. This finding reduced even further the credibility of the evidence in favour of spontaneous generation; and Pasteur was not slow to see that it also opened the way to a systematic analysis of the chemical requirements for microbial growth. This was an expectation that later generations amply fulfilled.

Pasteur's essay won the Alhumbert prize. A case has been made by some modern historians that this decision was unfair,[16] mainly on the ground that the adjudicators appointed by the Académie were biased. This contention is, in my view, a gross oversimplification of what actually went on. To begin with, the editors of the *Comptes rendus*, the Académie's journal, had allowed Pouchet extra space to ensure that his replies to criticism were published in their entirety. Then, there was no reason why spontaneous generation should have been chosen as the subject for the Alhumbert prize unless the Académie was anxious to give both sides of the argument a fair hearing. Given the public interest that had been aroused by Pouchet's claims, no other course of action would have been acceptable. This is confirmed by the punctilious wording used by the Académie in setting the subject for the prize: 'To endeavour, by means of well controlled experiments (expériences bien faites) to shed new light on the question of spontaneous generation'. To stress that the essay had to be based on 'well controlled experiments' indicates that the Académie, whatever preconceived views many of its members might have had on the subject, still harboured doubts about the experimental data that had so far been presented on both sides of the argument.

To turn now to the committee appointed by the Académie to judge the entries for the prize. Initially it had five members: Geoffroy St. Hilaire, Milne-Edwards, Serres, a former president of the Académie, Brogniart, and Flourens. Geoffroy St. Hilaire died before the judgement was made. He and Serres were then replaced by Claude Bernard and Coste. Flourens was the *rapporteur*, or secretary of the committee. He had been *Secrétaire perpétuel* of the Académie since 1833 and would therefore have been the normal choice for the position of *rapporteur*. I have already given a brief resumé of the relevant qualifications of Milne-Edwards and Claude Bernard. Brogniart was a distinguished botanist who, more than anyone else, had laid the foundations for the scientific study of fossil plants. He was co-editor of the botanical section of the *Comptes rendus*, so, with Milne-Edwards, who was the editor of the zoological section, both divisions of the biological membership of the Académie were represented. And Coste was the discoverer of the germinal vesicle in the mammalian egg. All the adjudicators for the Alhumbert prize were thus eminent scientific elder statesmen. It must be admitted that none of them was a believer in spontaneous generation. But it is doubtful whether believers of comparable eminence and seniority could have been found among the members of the Académie. The fact that Milne-Edwards, introducing his criticism of Pouchet, had found it necessary to apologize for raising the question of spontaneous generation at all indicates that disbelief must have been quite general.

In the event, Pouchet withdrew his entry for the prize because some of the adjudicators had apparently made their decision public even before reading the entries. There must have been some substance in Pouchet's complaint, for the Académie, sensitive to the accusation that it was not impartial, assembled a second panel of judges, Claude Bernard and Coste now being replaced by Dumas and Balard. Dumas, as I have mentioned, was a highly respected chemist who eventually himself became the *Secrétaire perpétuel*; and Balard was also a chemist, noted, among other things, for his discovery of the element bromium. Pouchet, still finding the committee unacceptably biased, again withdrew his entry. But it did not actually much matter whether the adjudicators believed in spontaneous generation or not. The remit of the committee was not to make a judgement about the existence or non-existence of spontaneous generation. They had been instructed to award the prize to the essay that most convincingly threw new light on the subject by means of 'well controlled experiments'. No plausible committee could possibly have given the prize to Pouchet in preference to Pasteur. Pasteur's essay was an experimental masterpiece, and one thing that Pouchet's experiments were not was 'well controlled'. Just how thoughtless Pouchet could be in the design of his experiments is well illustrated in three further reports that he communicated to the *Comptes rendus*.

In the first of these[17] Pouchet attempts to estimate the concentration of particles in atmospheric air by measuring the number deposited by falling snow. He chose an elevated site in Rouen where the snow was falling gently in large floccules. Then, over an area of about four square metres, he collected the surface layer of the fallen snow to a depth of about five centimetres. The snow was transferred to large glass basins and, when it had thawed, the number of recognizable living organisms in the water released was determined by microscopic examination of samples taken at various depths. Pouchet found that, among various kinds of debris, there were a few living organisms, but nothing like enough to account for the 'prodigious profusion' seen in his experiments on spontaneous generation. In a withering footnote in the *Annales* paper, Pasteur comments that he has not worked with snow, but it seems to him that it is the *lower* layer of the snow that should be examined, as it is this that sweeps the air clean.

The second report[18] describes an instrument that Pouchet has designed for the direct measurement of the number of organic particles in air. This instrument, which he calls an 'aeroscope', is essentially a glass tube with a funnel sealed tightly into one end; a smaller tube, connected to an aspirator, is sealed into the other end. The nozzle of the funnel, pointing to the inside of the tube, is aimed at the centre of a flat glass disc placed transversely across the lumen. The idea was that when air was aspirated into the tube it would deposit its load of particles onto the disc which could then be taken out and examined under the microscope. The concentration of particles in the air could be calculated from the volume of air aspirated. Pouchet did not see, in a volume of 1000 decilitres of air, a single infusorial egg or a single spore. A moment's thought might have led Pouchet to suspect that the stream of air directed onto the disc was more likely to dislodge particles rather than deposit them. But this seems never to have occurred to him. Indeed, he suggests, but only as an alternative, that the disc could be coated with adhesive, but he gives no details of the adhesive to be used and does not say whether the results differed from those obtained with uncoated discs. I can find no evidence of anyone else using the 'aeroscope'.

Pouchet's third report, written with his collaborators Joly and Musset,[19] appeared two years after Pasteur's *Annales* paper and is a direct challenge to Pasteur's experiments at high altitudes. The experiments of Pouchet and his colleagues were carried out in the high Pyrenees, at Recluse (2083 metres) and in a deep crevasse of the Maladetta Glacier (3000 metres). Flasks with boiled infusions were opened briefly at these sites and their contents later examined. Pouchet elaborates on the precautions he took with the flasks and claims that his precautions were even more stringent than those taken by Pasteur. When opened, all but one of the flasks teemed with microorganisms of various kinds, plant as well as animal. Although Pasteur, conscious of the limited number of flasks examined in his own experiments at high altitudes, had urged that they be repeated with larger numbers, Pouchet actually examined fewer flasks than Pasteur had done. The conclusion that Pouchet, Joly, and Musset drew from their collaborative experiments was that the teeming microbial populations seen in their flasks could not have been caused by aerial contamination, but were produced by spontaneous generation. Pouchet's peroration shows him to be utterly unmoved by Pasteur's work: 'Limited panspermia [the theory that the atmosphere is full of minute germs] does not exist, and heterogenesis [spontaneous generation], or the production of a new being devoid of parents but

formed out of the surrounding organic matter, is for us a reality.'
With his credibility as an experimentalist now so comprehen-
sively undermined, it seems unlikely that Pouchet's observations
at high altitudes would have brought him many new supporters
in the Académie.

Pouchet regularly used an infusion of hay as the organic solution with which he carried out his experiments on spontaneous
generation. It has been pointed out[20] that hay is regularly contaminated with hay bacilli (*Bacillus subtilis*), which, as Ferdinand
Cohn discovered in 1875,[21] produce spores under inclement
conditions. These spores resist boiling and would therefore still
have been viable in Pouchet's boiled hay infusions. The presence
of heat-resistant spores thus clouds all Pouchet's experiments, but
does not alone explain the results that he obtained. For Pouchet's
infusions produced a wide range of different organisms, and these,
of course, could not have come from hay bacillus spores alone.

A word about the Scientific Method, if there is one, and
Pasteur's alleged violation of its canons. No scientist who carries
out a series of exploratory experiments can fail to observe an
occasional aberrant result, one that cannot easily be accommo-
dated in the general trend of his measurements or that conflicts
with his preconceived ideas. All scientists, of course, have preconceived ideas (roughly what the Harvard historian Gerald
Holton[22] calls thematic propositions); no one undertakes a painstaking investigation with a completely open mind. The art of
science is to judge whether the aberrant result is trivial—a poor
measurement, a momentary fault in the apparatus, a clumsy
procedure—or whether the observation genuinely does run coun-
ter to the preconceived idea or perhaps even throws an unexpected new light on the problem as a whole. Good scientists
usually make the right judgement; bad scientists don't. Whether
to include the aberrant result in the condensed, edited, and
polished account of the work that is eventually published is also
a matter of judgement and does indeed involve the question of
probity. No editor, then or now, would agree to publish raw data.
But omission of data from what is finally published merges into
falsification only if the conclusion drawn from the work is wrong.
If that conclusion is right, the omission is of little interest to
anyone but the professional historian scouring the record for

discrepancies between the laboratory notebook and the finished paper. And on the question of spontaneous generation, Pasteur was almost always right.

Notes

1. H. Milne-Edwards (1859), *Compt. rend.* **48**: 23.
2. F. A. Pouchet (1858), *Compt. rend.* **47**: 979.
3. See Chapter 3, note 10; J. Farley and G. L. Geison (1974), *Bull. Hist. Med.* **48**: 161; G. L. Geison (1995), *The private science of Louis Pasteur.* Princeton University Press, N.J.; B. Latour (1984), *Les Microbes: guerre et paix, suivi de irréductions.* Paris; G. Pennetier (1907), *Un débat scientifique, Pouchet et Pasteur (1858–1868).* Paris.
4. See note 2.
5. See note 1.
6. F. Pouchet and M. Houzeau (1858), *Comp. rend.* **47**: 982.
7. See Chapter 3, note 11; Chapter 10, note 3.
8. F. A. Pouchet (1859), *Comp. rend.* **48**: 148.
9. F. A. Pouchet (1859), *Hétérogénie, ou traité de la génération spontanée.* Paris.
10. See note 3.
11. L. Pasteur (1860), *Compt. rend.* **50**: 303, 849; (1860), **51**: 348, 675; (1861), **52**: 16.
12. See Chapter 8, note 3.
13. L. Pasteur, Conférence faite aux soirées scientifiques de la Sorbonne 7 avril 1864, *Oeuvres de Pasteur* (1922–39) **2**: 238. Masson, Paris.
14. See note 3.
15. G. L. Geison, *The private science of Louis Pasteur.* See note 3.
16. See note 3.
17. F. A. Pouchet (1860), *Compt. rend.* **50**: 532.
18. F. A. Pouchet (1860), *Compt. rend.* **50**: 748.
19. F. A. Pouchet, N. Joly, and Ch. Musset (1863), *Compt. rend.* **57**: 558.
20. See Chapter 3, note 10.
21. F. Cohn (1875), *Beiträge zur Biologie der Pflanzen* **1**: 141.
22. G. Holton (1978), *The scientific imagination.* Cambridge University Press, Cambridge.

CHAPTER 11

Noxious particles

It was one thing to show that atmospheric air teemed with particles; it was another to show that particles did any harm. While this question does not bear directly on the credibility of spontaneous generation, its resolution changed the climate of opinion. Explanations of disease states in terms of spontaneous processes, whether generative or degenerative, became progressively less convincing and eventually unacceptable. Many scientists, over well-nigh half a century, contributed to this shift of opinion, but the prime mover was Agostino Bassi (1773–1856) (Fig. 11.1).

Bassi was a twin born at Mairago, a village in an agricultural region close to Lodi. He was educated at the gymnasium in Regia Città and at the University of Pavia, where he studied law and, part-time, some branches of natural history and medicine. When the Napoleonic armies entered Italy, he was appointed provincial administrator of Lodi, but his civil service career was interrupted by a depression in the economy which forced him, in order to maintain his family, to devote himself entirely to agricultural speculation. His interest in the fine wool produced by merino sheep led to the publication of a popular book on sheep husbandry. In 1808 he was made administrator of the hospitals in Lodi, but this did not put an end to his agricultural activities. He was deeply involved in public debate about the manufacture of cheese, and, over a period of a few years, published treatises on the cultivation of potatoes and the production of wine. In 1815 he contracted a severe eye disease that, for some time, deprived him almost entirely of his sight. He was appointed to the chair of general and special history (*Storia universale e particolare*) in the Istituto Filosofico di Lodi in 1824, but after his eye affliction he gave up civil service work and committed himself entirely to agricultural matters on his father's estate in Mairago.

Fig. 11.1 Agostino Bassi (1773–1856) wearing the insignia of the French Légion d'Honneur and the gold medal he received from the Emperor of Austria

The production of silk was a major industry in nineteenth-century Italy and depended, naturally, on the health of the silk-worm population. But this was always threatened by diseases of one sort or another and in particular by what is known in English as 'silk-worm rot'. The disease had many names in demotic Italian speech: *mal del segno, calcino, calcinaccio, moscardino*. The diseased worm stops eating; loses its mobility; curls up; develops a covering of powdery, white material; and eventually dies. Bassi began his enquiry into the cause of this disease in 1807 and worked on the problem continually for the next 20 years or more. By 1826, he had identified the causative agent unequivocally and amassed a great deal of information about it. But he chose not to publish his findings in the hope that they might bring him sub-stantial financial returns. Eventually he submitted an account of his investigations to a panel of professors at the University of Pavia and, receiving a favourable verdict, decided finally to pub-lish the work. It bore the title: On the silk-worm rot, a disease that attacks silk-worms and how to get rid of it even from the most heavily infested caterpillars[1] (Fig. 11.2). Part 1 (Theory) appeared

DE LA

MUSCARDINE

(Maladie des Vers à soie),

DE SES PRINCIPES

ET DE SA MARCHE;

MOYENS DE LA RECONNAITRE, DE LA PRÉVENIR
ET DE LA DETRUIRE.

Abrégé de l'ouvrage

DE M. LE DOCTEUR AGOSTINO BASSI
DE LODI;

Par M. le Comte Jacques Barbô,
DE MILAN.

Cet ouvrage, vendu au profit de l'auteur M. Bassi, est publié par les soins et aux frais de M. le comte J. Barbô, et déposé au bureau de l'Écho du monde savant, rue Guénégaud, 17, à Paris.

Paris,

CHEZ LES PRINCIPAUX LIBRAIRES.

—

1836

Fig. 11.2 Title-page of the French abridgement of Bassi's work on 'muscardine', the silk-rot

in 1835; Part 2 (Practice) in 1836. The dramatic, and completely unexpected, conclusion that Bassi drew from his observations was that silk-worm rot 'is always caused by a living organic being, a plant, a member of the cryptogam family, a parasitic fungus'.

This discovery was celebrated at two levels. Agriculturalists could not help but be excited by a discovery that identified the cause and promised the eradication of a major scourge of the silk industry. But at the level of fundamental biological principles, the discovery was even more sensational. It had, of course, been known since time immemorial that there were plants that parasitized other plants, but that there were plants that could parasitize a living animal, and kill it, was an entirely new concept. Its implications were profound. For medicine, it meant that infectious diseases might, at least in some cases, be due to the transmission of organisms that were barely visible or, even more difficult to detect, due to their air-borne seeds. For the theory of spontaneous generation, it meant that the area of obscurity in which vitalistic forces were still held to be operative was further drastically reduced. In the countless experiments that Bassi conducted before his work was published, the mode of transmission of the fungus was a recurrent theme. He examined infectivity in different parts of the diseased silk-worm and at different stages of the disease process. He studied the influence of temperature and of humidity on the efficiency of transmission. His papers of 1835 and 1836 precipitated a torrent of derivative investigations by other workers including some expert taxonomists. One, Giuseppe Crivelli, identified Bassi's fungus as a species of *Botrytis* and initially named it *B. paradoxa* but later changed the name to *B. Bassiana*. (Fig. 11.3).

Many honours followed in quick succession, from Italy, Austria, France, Germany, and, a little later, Russia. He was made a *chevalier de la légion d'honneur* and received a gold medal from the Emperor of Austria. But Bassi did not rest on his laurels, nor were his investigations limited to silk-worms. The publications that followed soon indicated that he regarded his findings on silk-worm rot not as a remarkable peculiarity, but as an example of a quite general process that underlay the aetiology of all infectious diseases, including those of humans. In 1846 his *Discourses on the nature and cure of pellagra* (*Discorsi sulla natura e cura della*

1. Ver à soie dans son état naturel.

2. Chenille processionnnelle à laquelle on a inoculé la muscardine.

3. Ver à soie mort de la muscardine.

4. Ver à soie couvert de l'efflorescence.

Fig. 11.3 Bassi's drawings of the progression of silk-rot in the caterpillar. The last drawing shows the emergence of the fungus

Pellagra) appeared. This is now known not to be an infectious disease, but its prevalence in areas where maize is a staple food and its symptomatology, which includes the deterioration of the skin and, often, diarrhoea, must have convinced him that it was. It is therefore not surprising that he made little headway in the analysis of this condition, but the attempt does indicate how far he had extrapolated the conclusions that he drew from his findings on silk-worms. The aetiology of pellagra was not, in the event, elucidated until well into the twentieth century. At the age of 78, in 1851, Bassi published *On the parasites that cause infectious diseases and the appropriate remedies* (*Dei parassiti generatori dei contagi e rispettivi remedi).* It contains the memorable conclusion: 'it follows from the laws that govern the development and reproduction of animal and plant parasites that the infectious diseases of plants and animals, including man, are all produced by parasitic organisms.' In assessing Pasteur's contribution to the germ theory of disease, it should be borne in mind that Bassi had been dead for more than a decade before Pasteur initiated his own studies on silk-worms. It is a remarkable fact that only at the end of his life did Bassi come into possession of a microscope, one constructed by Amici, but by then he was almost blind. He died of apoplexy in his eighty-second year.

The man who first provided decisive evidence that a plant parasite could cause disease in humans was Johann Lucas Schönlein (1743–1864). He was at the time professor of medicine at the University of Zürich. He had been appointed to this position in 1833, but left in 1839 because, being a Catholic, he was not given citizenship in that Protestant commune. He moved to Berlin where, as professor of medicine and director of a clinic at the Charité hospital, he rapidly became a highly influential figure in the medical scene. While still in Zürich he managed to obtain, via Paris, some dead silk-worms from Lombardy, and with this material he was able to confirm the findings of Bassi. It is clear, and Schönlein acknowledges it, that it was Bassi's work that provided the impetus for his own discovery. There happened, at the time, to be a case of ring-worm in the hospital, and when Schönlein examined the lesion, he saw at once that it involved a fungus. (The case had been diagnosed as *porrigo bispinosa*, porrigo being a now obsolete term embracing a variety of eczematous skin conditions, especially those of the scalp. Other names then used for ring-worm lesions were tigna (tinea) favosa or favus.) Schönlein isolated and illustrated the parasitic fungus which is now known as *Achorion schönleinii*. Schönlein's work was reported in Müller's *Archiv* as a summary of the letters he wrote to the editor.[2] It is clear from this report that Schönlein, like Bassi before him, did not regard his finding as a curiosity but as an example of a quite general category, the fungal diseases of man. Indeed, he regarded these diseases as analogous to the fungal infections of higher plants and drew attention to the 'beautifiul work' of Franz Unger on that subject.

There is one comment in Schönlein's communication that throws an interesting light on the hermetic mentality of the medical profession at that time. He complains that doctors hadn't yet taken on board the full significance of Bassi's work. His own discovery did indeed prompt a flurry of similar observations by others. Julius Vogel (1814–80), professor of medicine at Halle and author of the first atlas of pathological histology, identified the yeast-like fungus that caused thrush (*Candida albicans*); and David Gruby (1810–98), a Hungarian physician, demonstrated that *Trichophyton tonsurans* was the organism responsible for one form of patchy baldness (alopecia areata). These, and similar, observations rapidly established mycology as a discipline in its own right, but the medical profession as a whole simply did not envisage the possibility that *all* infectious diseases might be due to parasitic organisms, as Bassi had postulated. The conversion of medical opinion on this score was eventually brought about by the patient labour of Pasteur, Koch, and their associates, but the process took well-nigh a quarter of a century.

The story of the 'microbe hunters'—Pasteur, Robert Koch (1843–1910) (Fig. 11.4), and their disciples—has been told so often, in scholarly works, popular narratives, novels, films, and television programmes, that another account, however brief, would seem to be superfluous. But there are some aspects of this great saga that perhaps warrant a comment or two in the present context. Pasteur began his study of infection with an investigation of the diseases of silk-worms. This he did at the instigation of Dumas, who was much concerned at the devastation of the silk industry in the South of France by epidemics that had all but wiped out the silk-worm population. The two diseases that formed the centre of Pasteur's enquiry were 'pébrine', sometimes called 'gattine', and 'flacherie'. The name 'pébrine' is derived from the Provençal word for pepper (pebre) because, at one stage of the disease, the silk-worms have a mottled appearance said to resemble a sprinkling of pepper. Flacherie (flaccidity) is self-explanatory. Pasteur's investigation of these diseases was a direct extension of the work of Bassi, and in his 'Studies on the disease of silk-worms' (1870),[3] he acknowledged the fact. But, of course, between experiments done with the naked eye in 1807 and those done with an excellent microscope in the 1860s there is an

Fig. 11.4 Robert Koch (1843–1910)

enormous methodological gulf. Pasteur was able to resolve organisms that Bassi could not have seen and to study their behaviour with a precision that Bassi could only have imagined. The agent that gave rise to pébrine was found to be a minute protozoan that could be transmitted by contact, proximity, or vertically by passage through the egg. In the case of flacherie, a vibrio (a curved, usually motile, microorganism) was apparently involved. These discoveries did not represent a conceptual advance on the work of Bassi, but they provided evidence that organisms that could be detected only with the microscope (the smallest of Leeuwenhoek's animalcules) could also be parasites of animals and cause disease. Pasteur recounts how his thinking was dominated by the results of his work on spontaneous generation. Wherever he had found putrefaction or fermentation, in wine, beer, vinegar, blood, or urine, he had always found that living organisms were in some way involved. And nowhere could he find any convincing evidence that these organisms were generated spontaneously. It was thus a natural consequence of his observations to envisage the possibility, indeed probability, that infectious diseases would similarly be caused by parasitic microorganisms. After silk-worms, Pasteur extended his investigations to diseases in larger animals, but still animals of agricultural importance. He showed that a condition known as chicken cholera was also caused by a parasitic microorganism. He then turned his attention to anthrax. Anthrax is normally a disease of sheep and cattle, but it does occasionally pass to man, where it may produce lesions that range from a pustule on the skin to a fatal pneumonia. Pasteur, with characteristic showmanship, demonstrated that the disease was caused by the growth within the tissues of a rod-like organism that we now call *Bacillus anthracis*. Anthrax was a watershed. For it was the first instance of a disease in man that was caused by a bacterium. The vagaries of anthrax infection in the field were finally explained by Koch's discovery that anthrax bacilli formed spores that could remain dormant in the soil for years. The identification of further human bacterial pathogens was largely the work of Koch and his school. Koch, whose experimental prowess outstripped even that of Pasteur, himself identified the organisms responsible for tuberculosis and cholera. By the end of the century the causative agents of almost twenty of the infectious diseases of man had been identified.

These discoveries could hardly fail to influence the assumptions that underlay medical practice. But it was a slow process. Joseph Lister (1827–1912) introduced into surgery antiseptic procedures designed to exclude or destroy air-borne pathogens, but his methods were strongly opposed by many other practising surgeons. And even as late as 1892, during the great cholera epidemic in Hamburg,[4] Max Pettenkofer (1818–1921), an eminent physician and founder of a famous institute of hygiene in

Munich, was still arguing that the disease was caused by noxious miasmas generated by the foul conditions of life in the slums. When Robert Koch was asked by the authorities in Berlin to investigate the problem in Hamburg, his activities were relentlessly opposed by Pettenkofer.

Nonetheless, increasing numbers of medical practitioners were gradually won over to the views that had their origin in the work of Bassi and which were developed with such virtuosity by Pasteur, Koch, and their followers. As medicine entered the twentieth century, it came to be generally accepted that the atmosphere was indeed teeming with living microorganisms; that infectious processes, whether generative or degenerative, were not the product of intrinsic 'vital forces' or of imbalances in these forces; that the experimental evidence in favour of spontaneous generation was hopelessly defective. And what the conservative medical profession came to accept was soon accepted by the learned world as a whole. But there were individuals who still remained unconvinced. One of the most intransigent was Henry Charlton Bastian (1837–1915), professor of pathological anatomy at University College, London, whose activities I discuss in the next chapter. His defence of spontaneous generation throws an interesting light on the flux of scientific opinion at the time.

Notes

1. A. Bassi (1835–6), *Del mal del segno, calcinaccio o moscardino che afflige i bachi da seta e sul modo di liberarne le bigattaje anche le più infestate*. Lodi.
2. J. L. Schönlein (1839), *Arch. Anat. Physiol. wiss. Med.*, 85.
3. L. Pasteur (1870), 'Études sur la maladie des vers à soie', *Oeuvres de Pasteur* **4**: 5. Masson, Paris (1922–33).
4. R. J. Evans (1987), *Death in Hamburg*. Clarendon Press, Oxford.

CHAPTER 12

An English Pouchet

Bastian began his crusade in defence of spontaneous generation with three papers published together in *Nature* in 1870.[1]
The title of the series, 'Facts and reasonings concerning the
heterogenous evolution of living things', is itself of some significance, for, at a later stage, Bastian accused Pasteur not of defective experimentation, but of defective reasoning. The first paper
begins with a historical review which, like most of the reviews that
introduce papers on spontaneous generation, is tendentious. One
reading of the record shows that, from the very beginning, Bastian's mind was made up. He treats the experiments of Needham
and those of Spallanzani as of equal weight and concludes that
they simply neutralize each other. How anyone who had read
Spallanzani's painstaking monograph could consider his work no
better than Needham's is difficult to comprehend. So many of the
results reported by Spallanzani are ignored by Bastian that one
cannot help wondering whether Bastian had read Spallanzani's
monograph at all.

The historical review is followed by a discussion of the temperature required to destroy life. Bastian is heavily committed to
a lower range of temperatures and claims that 100 °C will kill
all infusoria. He ignores, or at least does not take seriously, the
evidence provided by other investigators that some organisms
easily survive this temperature. He suggests that there is an
analogy between the spontaneous generation of living forms and
the process of crystallization, an obsolete notion demolished by
Remak at least twenty years previously. That Bastian could make
such a suggestion in 1870 again raises doubts about whether he
had actually read the earlier literature. The centrepiece of the
paper is an extensive and extremely precise morphological
description of spontaneous generation as Bastian sees it under

the microscope. He describes a 'proligerous pellicle' in which unicellular organisms are formed by coacervation of granular material and its enclosure within a membrane. Amoebae are said to be formed from smaller rounded bodies, and fungal spores by the partitioning of granular masses. Many of these observations are illustrated by detailed drawings. It will surprise no modern reader that all this is pure fiction, but, in seeking an explanation for the reception that Bastian's work was given by his contemporaries, it is important to note that by 1870 the spontaneous generation of cells, as proposed by Schleiden and Schwann, had few adherents; and no other microscopist had seen the phenomena that Bastian described in such detail.

The experimental section of this paper introduces no new technical developments. The organic solution to be examined is again an infusion of hay contained within a flask whose neck has been drawn out to a narrow orifice. The infusion is boiled until it froths over and the air is expelled from the flask. The neck is then sealed with a blow-torch. Bastian finds growth of microorganisms in flasks handled in this way. Contemporaries familiar with the literature on spontaneous generation would have known that some microorganisms did indeed survive temperatures higher than that of boiling water and that hay was a particularly unfortunate choice of organic material. They would also have known that this experiment had been conducted repeatedly with other infusions for more than a century and had, in the great majority of cases, produced no growth of microorganisms. John Tyndall, of whom more later, added a further objection. He pointed out that boiling the solution until it frothed would cause 'bumping', which might deposit organisms on the cooler parts of the flask and thus permit their survival. Bastian's pre-emptive answer to these objections was to invoke a logical rule which, as I have discussed previously, is wholly inapplicable to the experimental situation: 'A single positive result', he argues, 'when Schwann's experiment has been legitimately performed, is of far more importance towards the settlement of the question in dispute than five hundred negative results.' And he quotes three authors, little known in the context of spontaneous generation, who had obtained growth of microorganisms in solutions heated to much higher temperatures and for much longer periods.

In his second paper Bastian investigates microbial growth in a wide range of different solutions, and, although he does not regard it as necessary, he examines the effect of higher temperatures. To begin with, he boils his solutions at $100\,^{\circ}$C for 10–20 minutes. He uses neutral, slightly alkaline, or slightly acid infusions (filtered beef juice, a 'decoction' of beef and carrot, a solution of beef, carrot and turnip, various infusions of turnip at different strengths, and an infusion of hay). In only one of eight experiments does he fail to observe microbial growth. He is able to identify and illustrate some of the organisms (bacteria, 'monads', and moulds), but he also describes 'irregular moving shapes, granules, protoplasmic masses and protoplasmic-looking material'. He then turns to Pasteur's observation that microbial growth can occur in inorganic solutions. He finds nothing extraordinary in this for, in his view, inorganic solutions can readily produce organic solutions and these, in turn, can readily produce colloids. Colloids, he argues, are the key step in the process that generates living forms: 'The colloid possesses ENERGIA. It may be looked upon as the probable primary source of the force appearing in the phenomena of vitality.' He describes eight similar experiments with solutions of defined composition (they are not all inorganic by modern criteria). Again, in only one experiment does he fail to observe microbial growth.

The next section of this paper deals with the effects of higher temperatures. The various inorganic and organic solutions are now heated in a 'wrought iron digester' for 4 hours at a temperature of $146\,^{\circ}$–$153\,^{\circ}$C. In all of four flasks containing the solutions Bastian finds extensive microbial growth. These experiments were conducted in collaboration with 'Dr. Frankland' (a distinguished chemist); one of the flasks was opened in the presence of 'Dr. Sharpey', another in the presence of 'Professor Huxley'.

Bastian's third paper begins with further comments on Frankland's experiments. Fungi and their spores, Bastian claims, are killed by a few minutes' boiling and are reduced to debris by the temperatures reached in the digester. To check that his simple solutions do not contain germs or spores, he examines them under the microscope but finds nothing except, in one case, a solution of ammonium tartrate. However, in a more recent batch of the tartrate, obtained from the manufacturer, he finds no organisms. Examining the older contaminated batch, he contends that he can see the life-forms beginning to develop within crystals. First he sees a 'primordial monad' which develops into a bacterium; this is then transformed into a vibrio and finally into a *Leptothrix*. This process he terms 'isomeric transformation'. Other microscopists had, of course, long ago become suspicious of such apparent transformations.

The paper ends with a severe criticism of Pasteur's experiments. Bastian admits that Pasteur is an excellent chemist but doubts whether the non-existence of spontaneous generation had been 'mathématiquement démontré', as Pasteur had claimed. What is more, Bastian accuses Pasteur of ignoring conflicting results (which is probably true) and criticizes his experiments with milk on the ground that the effect of acidity or alkalinity on the generation of life had not been adequately controlled. 'Unfortunately for the cause of Truth,' he continues, 'people have been so blinded by his skill and precision as a mere experimenter, that only too many have failed to discover his shortcomings as a reasoner.' Bastian sums up his own position in the following words:

> On account of this *a priori* probability, and in the face of this evidence, I am, therefore, content, and as I think justified, in believing that living things may and do arise *de novo*. Such a belief necessarily carries with it a rejection of M. Pasteur's Theory of Putrefaction, and the so-called 'Germ theory of Disease'.

Bastian's findings were soon challenged by William Roberts, a highly respected Manchester physician, briefly in a letter to *Nature* in 1873[2] and more exhaustively in a communication to the Royal Society in 1874.[3] The latter begins with a variant of Pasteur's 'swan-necked' flask experiment. A narrow tube, recurved at its lower end, is inserted via a cotton wool barrier into a flask containing the organic infusion. If the recurved end of the tube is kept clear of the boiling infusion, no microbial growth occurs even when the other end is left open to the air. But if the recurved end is lowered into the infusion, then fermentation and microbial growth do occur. Roberts concludes from this experiment, much

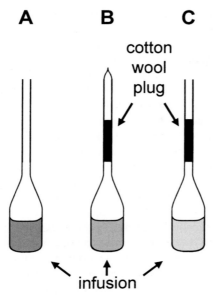

Fig. 12.1 The plugged pipettes used by Roberts. (A) is open to the air. (B) is plugged and sealed. (C) is plugged but not sealed (see text)

as Pasteur had done, that the recurved end of the tube traps the particles present in the air and thus shields the infusion from contamination.

The second part of Roberts's paper introduces a procedure that does not embody any new experimental principle, but which greatly facilitates experiments with open or sealed vessels. Bulb pipettes are sealed at one end to provide a chamber for the organic infusion (Fig. 12.1). The other end of the pipette is either left open (A), or plugged with cotton wool and then sealed (B), or plugged with cotton wool and sealed but then opened to the air by removal of the seal (C). The ease of this procedure permitted Roberts to do hundreds of experiments in which the effect of filtered and unfiltered air could be compared. A wide range of organic infusions was examined and a special study made of the temperatures required to kill different microorganisms.

Roberts reports that his results do not agree with those of Bastian. Boiling the infusion for a few minutes does not necessarily sterilize it. In order to kill some microorganisms, much higher temperatures are required, and both the degree and the duration of the heat applied are of importance. Slightly alkaline fluids are more difficult to sterilize, and more than an hour is required to sterilize alkalinized hay infusions. Fungi and yeasts are never produced in infusions boiled for even a brief period unless the infusions are contaminated by particles in the surrounding air. In a table that summarizes his results, Roberts makes a point of including those that support the theory of spontaneous generation and those that argue against it. Out of 90 experiments in which the organic solution was heated and then shielded from the outside air, 67 produced no microbial growth but 23 did. Roberts admits that the main aim of his experiments was to confirm 'in the fullest manner the main propositions of the "panspermic" theory', and he concludes that the 'relatively few discordant results are not strong enough to allow deduction of "abiogenesis" [spontaneous generation] but impose limits on it. Even if cases of "abiogenesis" did occur this would not overturn the "panspermic" theory.' This is a very temperate conclusion that contrasts strongly with the dogmatism that pervades much of the writing on the subject of spontaneous generation.

Papers by Bastian appeared in various journals in each of the years that followed the publication of his *Nature* trio, but it was not until 1876 that he elicited a reply from Pasteur. In that year, on 15 June, Bastian presented the results of his more recent research at a meeting of the Royal Society. An account of the meeting appeared in *Nature* soon thereafter.[4] Bastian's communication concerns fermentation of urine, a subject considered in detail in Pasteur's *Annales* paper. Bastian claims that boiled urine shielded from the air can undergo fermentation and produce microorganisms if it is rendered neutral or slightly alkaline by the addition of a solution of potash (liquor potassiae). The amount of potash to be added is critical, for, if too much is added, the urine remains sterile. In some cases, Bastian finds that fermentation of the boiled urine can be induced by raising the temperature at which it is subsequently incubated. Incubation at 37 °C might not be hot enough, and temperatures as high as 50 °C are sometimes required. The experiment is conducted with the now familiar flask plugged with a barrier of cotton wool. A sealed capillary tube containing the liquor potassiae is immersed in boiling water for up to 15 minutes and then inserted into the flask via the cotton wool barrier. The urine in the flask is then boiled and the liquor potassiae added simply by shaking the flask and breaking the capillary tube.

Bastian considers the objections that might be made to his experiments. First, the liquor potassiae might contain viable germs. But if this were so, a few drops of the fluid would be enough to induce fermentation, whereas he finds that a critical amount of the potash solution is required. Secondly, the potash solution might reactivate dormant germs that are still present in the boiled urine. His answer to this is that germs are killed by boiling in acid solutions like urine; he concedes that bacterial growth occurs readily in high concentrations of potassium but claims that only a limited range of alkalinity allows the process to be initiated. The conclusion that he draws from these experiments is that 'the fertilizing agent [liquor potassiae] acts by helping to initiate chemical changes of a fermentative character in a fluid devoid of living organisms or living germs'. This process he now calls 'archebiosis'.

Bastian must have been spoiling for a fight for on 10 July, about three weeks after his communication to the Royal Society, he sent a short note to the Académie des sciences in Paris setting out his position. Pasteur replied at the next meeting of the Académie and his response appeared in the *Comptes rendus* dated 17 July. Pasteur's reply is polite and very temperate. He had not apparently read Bastian's earlier papers and limits his comments to the latter's experiments on the fermentation of urine. Pasteur makes it clear that he does not doubt that Bastian had given an accurate description of what he had observed, but he cannot agree with Bastian's interpretation of these observations. Bastian later quotes the relevant passage: 'I hasten to declare that Dr. Bastian's experiments are indeed very precise; they do most commonly yield the results that he indicates . . . Between M. Bastian and myself there is therefore only a difference in the interpretation of experiments that we have both carried out.' But this difference in interpretation is at the heart of the matter, for Pasteur now gives it as his view that the growth of microorganisms in Bastian's experiments is due to contamination by germs surviving in the liquor potassiae. Bastian refers to this interpretation as an 'astounding hypothesis'. In its support Pasteur adduces evidence that temperatures higher than that of boiling water are required to sterilize the potash. A solution of potash must be heated to 110 °C and, if solid potash is used, it must be heated to redness.

Bastian's reply to Pasteur appeared in *Nature* in the same year.[5] It is entitled 'The fermentation of urine and the germ theory'. Bastian now considers that the essential question is whether germs can survive boiling in liquor potassiae (as defined by the British Pharmacopoeia). He tabulates what he sees as the main differences between himself and Pasteur, but he does not introduce any new experimental evidence. Against the view that his liquor potassiae is still contaminated after boiling, he again rehearses the arguments that he has already published.

At the end of that year William Roberts communicated to the Royal Society the results of his own experiments with a solution of potash. A report of this work appeared in *Nature* in the following year under the title 'The spontaneous generation question'.[6] The observations reported by Roberts do not at all agree with those of Bastian. Using a procedure very much like Bastian's, Roberts finds that if the potash solution is heated for long enough and at a high enough temperature, the boiled urine remains sterile. To ensure sterility throughout the manipulations, Roberts lets the boiled urine stand for 2 weeks before experimenting with it and heats the potash solution to 280 °F. The flasks are incubated at 70–80 °F for a fortnight and then at 122 °F for 3 days, Bastian having claimed that a higher temperature sometimes facilitates the process of spontaneous generation. Roberts examines a wide range of different organic solutions including infusions of hay; but he finds, unlike Bastian, that they all remain sterile if proper precautions are taken to make sure that the solutions are sterile in the first place. Roberts notes that alkaline solutions are harder to sterilize than acid ones and suggests that adding the potash to the urine in order to alkalinize it might permit pre-existing germs to survive. He evidently finds it necessary to point out to Bastian that a precipitate of inorganic phosphate is produced in urine and that this might be mistaken for bacterial growth. Roberts's paper has an addendum by Tyndall. Tyndall fully agrees with Roberts and reports that in similar experiments his flasks remained sterile for two months.

Nature, welcoming a public controversy on an important subject, published the exchanges between Bastian and Pasteur immediately after the paper by Roberts and Tyndall, and headed all these communications 'The spontaneous generation question'.[7] Pasteur's reply to Bastian was communicated to the Académie on 8 January 1877. It is a joint paper by Pasteur and Joubert. Nine years previously Pasteur had suffered a stroke that rendered the left side of his body largely paralysed and his left hand irreversibly contracted. All his work after the stroke was done through the hands of collaborators. Pasteur again stresses the advantages of solid potash for this kind of experiment, partly because it is the only true potash and partly because it enables him to neutralize the urine exactly. He has, however, carried out experiments in which the urine is saturated with potash, and these show that microorganisms can grow even in the saturated solution. If liquor potassiae is to be used, it must be heated to 110 °C. If this is done and appropriate precautions are taken, the boiled urine will remain sterile. Bastian replies that he has never used solid potash and, in doing so, Pasteur is defecting needlessly from the procedure that Bastian had described. Bastian also finds it hard to believe that germs could survive boiling in a solution as caustic as liquor potassiae. If the boiled liquor potassiae had still contained microorganisms, a few drops would have been enough to fertilize the urine, whereas, Bastian maintains, only a very narrow range of added potash can bring this about. Bastian ends his reply with the remark that he does not believe there is much point in debating this matter with Pasteur any further.

Nonetheless, Pasteur did continue with the debate. On 29 January he made a further communication to the Académie. Now Pasteur's tone becomes more assertive and his remarks are tinged with irony. He refers to Bastian in one place as 'the learned London professor of pathological anatomy' and in another as 'the English savant'. He again defends his use of solid potash because it is free of any organic matter, but liquor potassiae, whether it is made up in accordance with the British 'or any other' pharmacopoeia, would be acceptable provided it is heated to 110 °C for 20 minutes or 130 °C for 5 minutes. Bastian replies that, in the light of Pasteur's challenge, he has redone his experiments with liquor potassiae now heated to 110 °C for 60 minutes, but the results are unchanged. Fermentation takes place in all flasks.

At this point Pasteur asked the Académie to appoint a commission to report on the fact that was under discussion between himself and Bastian. What ensued is an echo of the events that took place more than fifteen years earlier when Pouchet withdrew his entry for the Alhumbert prize. Regrettably, the only account of the proceedings is that published in *Nature* by Bastian, and on the strength of it some historians have again come to the conclusion that the commission of the Académie was biased. It is likely that the position was again more complex than might appear from Bastian's account. The Académie appointed only three adjudicators. This, in itself, probably has some significance, for had it been thought by the members of the Académie that a matter of genuine importance was to be considered, then a more representative commission would have been assembled. The three adjudicators were Dumas, now the *Secrétaire perpétuel* and the obvious *rapporteur*, Milne-Edwards for the animal biologists, and Boussingault for the plant biologists. I have already listed Milne-Edwards's qualifications; Jean-Baptiste Boussingault was a distinguished chemist with a special interest in plant physiology. The terms of reference of the commission were that they were to express an opinion on the fact that was under discussion between Dr. Bastian and M. Pasteur.

Bastian thought it worthwhile to publish his version of what went on in Paris. His account appeared in *Nature* and was entitled 'The commission of the French Academy and the Pasteur–Bastian experiments'.[8] Bastian takes up Pasteur's challenge 'on the sole condition that the solution [of liquor potassiae] shall be raised beforehand to 110° for 20 minutes, or 130° for five minutes'. In other respects Bastian requires that the experimental procedure be that with which he had previously demonstrated spontaneous generation. Bastian wrote to the commission on 27 February 1877 in the expectation that the commission would not want to 'express an opinion' without having the two protagonists perform their experiments. He therefore asks the Académie to fix a convenient date. This request apparently remained unanswered for some time. Dumas had in fact replied to Bastian on 5 May, but the letter was delayed in the post because Bastian had changed his address and the letter had had to be re-addressed. In his reply, Dumas agreed to Bastian's conditions and offered a laboratory at the Ecole normale, but he was prepared to accept any other site that Bastian might prefer. Dumas wrote again on 18 May, enclosing a duplicate of his earlier letter. He now informs Bastian that Pasteur has offered his own laboratory, but repeats that any site that Bastian chooses would be acceptable. 14 July would be a convenient date for Pasteur and the commission; and since the commission was already familiar with Pasteur's experiments, it is only Bastian's that it would wish to see.

A further missive from Bastian insists that the judgement be restricted to the fermentation of urine alkalinized by liquor potassiae heated to 110 °C for 20 minutes 'at least'. He declares himself unwilling to come if the commission proposes to express an opinion on the germ theory of disease or on spontaneous generation. If the commission were to insist on further experiments, the matter might be dragged out indefinitely. He accepts 14 July. Another letter from Bastian. 14 July is now no longer convenient, and he suggests 15 July or thereabouts. Dumas agrees and again accepts all Bastian's conditions. But Bastian still wishes to be assured that the adjudication will be limited to the specific matter in question and that no new experiments will be demanded of him. He also suggests that a French friend of his might act as an interpreter. Dumas once more accepts Bastian's conditions and informs him that Milne-Edwards speaks English very well. Bastian finally arrives in Paris on 13 July.

Only Dumas and Milne-Edwards are present; Boussingault is unable to come for unstated domestic reasons. Milne-Edwards objects to Bastian's demand that there be no change in his experimental conditions. If these cannot be varied, Milne-Edwards is not prepared to take part in the commission. No doubt his attitude here was determined by his previous experience with Pouchet. He was well aware of the technical pitfalls that beset experiments of this kind and the ease with which the organic solution could be contaminated. He clearly did not wish to be implicated in passing judgement on what he was convinced would be a spurious result.

To replace Boussingault, Bastian suggests a number of names, but the Académie appoints Van Tieghem. The reason for this choice is not clear from the record, but it was, of course, the Académie's prerogative to appoint whomever it chose, and it was hardly Bastian's place to suggest who his judges should be. Nonetheless, Bastian cannot refrain from adding the comment that Van Tieghem had been a pupil of Pasteur's and could therefore be regarded as an opponent. Bastian then receives a letter from Van Tieghem fixing the time and place: Pasteur's laboratory at the Ecole normale at 8 a.m. on the following day. Van Tieghem will himself replace Pasteur. This substitution was, no doubt, necessary because Pasteur was by then half-paralysed, but Bastian does not mention this.

Milne-Edwards, Van Tieghem, Pasteur, and Bastian arrive at the appointed time, but Dumas has not yet appeared. Given that he was then 77 years old, his late arrival might perhaps be forgiven. Milne-Edwards explains that he had not been informed of all the correspondence between Bastian and Dumas, declares himself unwilling to participate in the commission under the conditions laid down by Bastian, and leaves. Van Tieghem also leaves, but apparently only to find out why Dumas has not yet arrived. Van Tieghem returns after an hour and stays to talk with Bastian in an upper room. When they descend they learn from Pasteur that Dumas has arrived. However, when Dumas hears that Milne-Edwards has left, he declares the commission at an end. He does not communicate any further with either Van Tieghem or Bastian.

There is no reason to doubt the facts recorded in Bastian's account, but one cannot help wondering about the facts that were not recorded. It must at once be admitted that the representatives of the Académie did indeed behave in an off-hand manner; that the behaviour of the aged Dumas, although initially punctilious, left much to be desired; that the commission appointed by the Académie should have had a larger membership. But it should also be remembered that the biological members of the Académie had long ago made up their minds about experiments that purported to show the generation of life from inanimate material. They were familiar with the evidence adduced by Pouchet and its comprehensive destruction by Pasteur. They may well have regarded Bastian simply as a crank who was wasting their time; and there is no doubt that he had made a thorough nuisance of himself. When Dumas learned that only one member of his appointed commission was now prepared to act, what else could he do but terminate the proceedings? All this is little more than conjecture, but it remains a pity that we have no account of the episode from the Académie's point of view. Bastian was not deterred. He was still writing about the spontaneous origin of living matter in 1905; but by that time his was a lonely voice.

Notes

1. H. C. Bastian (1870), *Nature, Lond.* **2**: 170, 193; (1870), **3**: 219.
2. W. Roberts (1873), *Nature, Lond.* **7**: 302.
3. W. Roberts (1874), *Phil. Trans. Roy. Soc.* **164**: 457.
4. H. C. Bastian (1876), *Nature, Lond.* **14**: 220.
5. H. C. Bastian (1876), *Nature, Lond.* **14**: 309.
6. W. Roberts (1877), *Nature, Lond.* **15**: 302.
7. (1877), *Nature, Lond.* **15**: 313, 380.
8. H. C. Bastian (1877), *Nature, Lond.* **16**: 277.

CHAPTER 13

Seeing the floating particles

John Tyndall (1820–93) (Fig. 13.1) was a remarkable Irishman who made many outstanding contributions to physics, but none more memorable than what we now eponymously call the 'Tyndall effect'. He was largely self-educated and, by dint of quite exceptional experimental skill, eventually became the professor of natural philosophy at the Royal Institution in London. He succeeded Michael Faraday as its superintendent. The Tyndall effect is produced by the scattering of light by particles in solution; if the light passes obliquely through a transparent solution, the particles suspended in it become visible. It had, of course, been known since time immemorial that a shaft of oblique light passing through a darkened room will illuminate the dust particles in the air; but in Tyndall's hands, light scattering was converted into a systematic analytical method. It remains a standard procedure to the present day. The seminal contribution that Tyndall made to the study of spontaneous generation was that the air-borne particles that contaminated experiments on that subject could now be seen.

Tyndall's preparatory experiments were reported in *Nature* under the title 'On floating matter and beams of light'.[1] A beam, transmitted longitudinally through a glass tube, illuminates masses of particles that reflect and refract the light. Tyndall describes the appearance as one of 'astonishing complexity and beauty'. The suspended particles are found to be very sensitive to external stimuli. An instant's contact with a spirit lamp causes a violent upward current; even the warmth of a finger is enough. The movement of the particles produces 'beautiful vortices', and there is also a dark current that Tyndall explains in terms of flow theory.

Fig. 13.1 John Tyndall (1820–93)

To examine the behaviour of the suspended particles more systematically, Tyndall now constructs a square chamber, the upper half of which is glazed to permit illumination and direct observation of the particles. The floor of the chamber is composed of transverse rails that support a thick pad of cotton wool, and in the roof there is a rose-burner enclosed within a brass chimney. When the burner is lit, the heated air escapes through the chimney and is replaced by cooler air filtered through the cotton wool. The filtered air does not produce uniform darkening of the chamber, but dark stream lines appear which throw the bright particles to the periphery.

With spontaneous generation clearly in mind, Tyndall now sets about destroying the particles and not merely redistributing them. In this case, a cubical glass shade serves as the vessel that can be illuminated and observed. The shade is traversed by a platinum wire connected to a battery. When the wire is heated to white heat for 5 minutes there is a marked diminution in the amount of illuminated floating matter within the glass shade. After 10 minutes the particles are totally destroyed. Tyndall's next step is to test the efficacy of a heated platinum tube. The platinum tube is hermetically sealed into one end of a larger glass tube in which the illuminated particles can be observed. The glass tube is then exhausted and re-filled with air that enters via the heated platinum tube. It is only when the platinum is heated to a bright redness and the air enters very slowly that the floating matter is totally destroyed.

Tyndall then confronts the question of spontaneous generation directly, but, in the first instance, only to 'clear the field'. He proposes to discuss the matter more thoroughly in a later paper. 'Clearing the field' involves a critical review of the experiments described by other workers, especially Pouchet. He has two objections to Pouchet's work. Pouchet had used water produced by the combustion of hydrogen in air, but Tyndall points out that in this process the water is obliged to trickle through air and thus collects air-borne particles. He shows by indirect illumination that water so produced is indeed full of them. Tyndall's second objection has already been mentioned: viable microorganisms in the organic infusion might escape destruction when the infusion is boiled, because 'bumping' deposits them on cooler parts of the flask. Pasteur's experiments with open flasks in cellars, where there is little movement of air, are now subjected to scrutiny. Tyndall is able to confirm them, but goes further. In a cellar the flasks still contain particles, although not as many as there are in the outside air. The 'tranquil' particle-free air is provided by the flasks themselves, for the air-borne dust is gradually deposited and clings firmly to their sides. The flasks thus do not need to be corked. If they are placed on a table and left undisturbed they eventually become optically empty. This self-cleaning process is monitored in a tall observation chamber meticulously sealed. Within a week it has become optically empty and the dust can be seen clinging to the interior surfaces of the chamber.

A few years later Tyndall published a more detailed examination of the spontaneous generation question.[2] Apparently a large body of medical opinion was still unsure whether 'disease-germs' were the only cause of infection or whether, in some cases, this might be caused by a process analogous to spontaneous generation. This indecision had recently found expression in the *British Medical Journal*, and Tyndall's paper was an attempt to settle the matter once and for all. His paper has three parts. The first describes still more elaborate observation chambers in which the growth of microorganisms can be monitored under the most exacting conditions; the second contains further technical criticisms of previous claims; and the third discusses the analogy between infectious disease and the microbial contamination of organic infusions. The observation chambers are still made of wood but have a glass front and glass panels inserted into the sides. The floor is now pierced with one or two rows of holes to admit test-tubes holding the organic solutions (Fig. 13.2). Two pieces of twisted glass tubing are inserted into the chamber through holes drilled in the roof. These are plugged with cotton wool to trap any particles in the air that might enter the chamber in response to temperature fluctuations. All junctions are made airtight and all crevices meticulously sealed. The whole of the interior is coated with glycerine to hold any particles adhering to the inner walls.

The experiments are conducted only when optical monitoring shows that the chamber has been completely cleared of particles (is 'mote' free). Some two dozen different infusions are tested, including extracts of animal and vegetable tissue, both acid and alkaline. The boiled organic infusion in 600 flasks exposed to the outside air all undergo putrefaction; but in no case is there any putrefaction in the infusions exposed to the mote-free air in the chambers. Tyndall stresses that air cleared of particles in this way is in no way 'tortured'. He concludes that putrefaction is indissolubly linked to particles in air.

His criticisms of claims to the contrary contain no biological theorization, but merely seek to identify undetected errors in

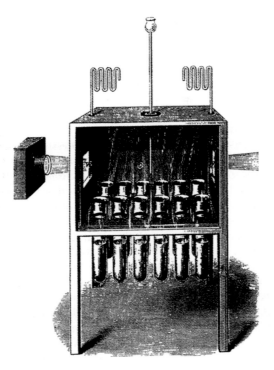

Fig. 13.2 One of the dust-free chambers constructed by Tyndall to examine the question of spontaneous generation

technique. With calcined air or air filtered through cotton wool, he finds no putrefaction in his boiled organic solutions if scrupulous attention is paid to the elimination of external contamination. In a bell jar traversed by a platinum wire, he again finds no putrefaction when the wire has been adequately heated and the air then pumped out. He repeats Bastian's experiment 139 times, but fails to see spontaneous generation of microorganisms in any of his flasks. This he demonstrates to the Royal Society. He can, however, produce growth of microorganisms if, like Bastian, he attempts to seal the narrowed neck of the flask while the solution is still boiling. He again attributes the results obtained by Bastian to viable organisms deposited on cooler parts of the flask by the process of bumping. He also suggests that the tip of Bastian's pipette might have been contaminated. The experiments of William Roberts with modified bulb pipettes win his praise, but he points out that untreated cotton wool always contains germs, which might account for the rather high proportion of cases in which Roberts observed microbial growth in the boiled infusion. His conclusion to this section of the paper is more sweeping: 'life has never been proved to appear independently of antecedent life.'

The third part of the paper describes experiments in which Tyndall attempts to mimic the process of natural infection under experimental conditions. One hundred holes are drilled into the bottom of the optical chamber to accommodate the tubes holding the organic infusions: 30 of hay, 35 of turnips, and 35 of beef. These are simply exposed to atmospheric air, and both the incidence of putrefaction and the timing of its onset monitored. Infection does not occur uniformly in all tubes. Often there is a lag period, and the putrefaction may vary in its character and in the rate of its progression. Tyndall concludes that the infectivity of the air (the *contagium vivum*) is not uniform, but is borne in clouds or packets containing particles at different densities and of different composition. This was the conclusion that Pasteur had also reached, although, of course, with much less sophisticated methods. Tyndall then sets up an experiment to test the 'fetid gas' or 'miasmatic' theory of infection. Tubes containing putrid infusions emitting a foul smell are inserted into the observation chamber near tubes in which the infusions are uncontaminated. Both are boiled and maintained in particle-free air. No cross-infection occurs despite the fetid odour, but if a drop of the putrid infusion is transferred to the clear infusion, putrefaction is at once initiated. Tyndall naturally draws an analogy between these findings and the spread of infectious diseases and has no hesitation in concluding that infectious diseases are caused by infectious particles.

In the following year Tyndall presented to the Royal Society an addendum to his previous paper.[3] This preliminary note deals with cases in which he had found it difficult to sterilize materials by the procedures that he normally used. In particular, he, like all before him, had had problems with infusions of hay. These he now explores in depth. He finds that hay mown the previous year is easily sterilized by 5 minutes' boiling, but, as the hay ages, it becomes increasingly difficult to sterilize it. A five-year-old sample that he has managed to obtain is the most resistant of all. So exhaustive were Tyndall's experiments with hay that the laboratories at the Royal Institution became hopelessly contaminated and new ones had to be constructed in Kew Gardens. In the end Tyndall succeeded in sterilizing all the materials he worked with and concluded that if the experiments were done with a scrupulous enough technique they provided no evidence at all to support spontaneous generation.

If there was anyone who administered a *coup mortel* to the idea of spontaneous generation, it was Tyndall rather than Pasteur. But it was another matter to convince those who believed in it. As mentioned earlier, Bastian was still arguing his case well into the twentieth century, and even after the Second World War O. B. Lepeshinskaya won the plaudits of Joseph Stalin for her alleged demonstration that cells and even tissues could be formed from unorganized matter.[4] In an ideology that had no room for divine intervention, the observation that living forms could be generated from inanimate material by natural processes was, of course, welcome. After Tyndall, however, experiments in support of spontaneous generation were rare, and there were none that scientists could take seriously. But perhaps the most remarkable feature of this long and acrimonious controversy is that not one of those who advanced experimental evidence in support of spontaneous generation changed his mind, or, at least, admitted to doing so. Perhaps Max Planck, the doyen of twentieth-century German physicists, was right after all when he concluded that 'a new scientific truth does not in general prevail because its opponents are convinced by it and declare themselves to be so, but rather because these opponents gradually pass away and the coming generation is intimately familiar with the truth from the beginning'.[5]

Notes

1. J. Tyndall (1870), *Nature, Lond.* **1**: 499.
2. J. Tyndall (1876), *Proc. Roy. Soc.* **24**: 171.
3. J. Tyndall (1877), *Proc. Roy. Soc.* **25**: 503.

4. See Z. A. Medvedev (1969), *The rise and fall of T. D. Lysenko.* Columbia University Press, N.Y.
5. M. Planck (1958), *Physikalische Abhandlungen und Vorträge* **3**: 389. Braunschweig.

CHAPTER 14

An epilogue about another subject

While it has come to be accepted that no experiment has ever been devised that can convincingly demonstrate the formation of life from inanimate material, how life originated in the first place remains an open question. For those who accept divine intervention and the miracles that may flow from it, there is no problem. But those who reject this easy option are faced with a quandary. For if there is no divine intervention, it is difficult to see an alternative to some form of molecular self-assembly. Some historians have therefore maintained that in reaching that conclusion we have come full circle. Having argued for centuries against the idea of spontaneous generation, scientists, it is alleged, are now arguing in its favour. This contention is a misconception that arises because the term spontaneous generation is here used to include two phenomena, one testable, the other untestable. What I have been concerned with in this book is the testable phenomenon. In the world that now exists, or that existed when scientists in the past conducted their experiments, is it possible or has it been possible to demonstrate the transmutation of inanimate matter into living forms? The answer is no. But in a different world, one not susceptible, or no longer susceptible, to experiments, no categorical answer can be given.

Scientists are not now trying to prove that spontaneous generation, as we have defined it, exists after all. What they are doing is exploring chemical reactions that might plausibly have taken place on the surface of the earth when life first made its appearance. In particular, they study those chemical reactions that might assemble simple molecules into more complex ones and eventually into molecules that we now know to be characteristic of life-forms. Almost without exception, scientists accept the geological and astronomical evidence that physical conditions on earth in the distant past were vastly different from what they are now, and scientists accept also that there can be no certainty about the details. But they do not in any case imagine that living organisms were originally generated from inanimate molecules in one abrupt step. They envisage, on the contrary, an extremely gradual process stretching over aeons of geological time and subject to the continuous pressure of Darwinian natural selection. This view of the origin of life has little in common with the historical concept that we have chosen to call spontaneous generation. Indeed, the only important similarity between the two is that neither requires the intervention of the supernatural.

In the period preceding the Second World War it was generally thought that proteins were the characteristic, and indispensable, components of living matter. Proteins were known to be able to catalyse chemical reactions, to form complex structures, and to undergo highly specific interactions with other proteins. The first modern attempts to unravel the origin of life in chemical terms were therefore centred on proteins, and how, under conditions thought to mimic those of the primitive earth, proteins could undergo self-assembly.[1] Reactions were discovered that permitted amino acids, the essential subunits of proteins, to be formed from much smaller compounds, especially those that might have been present in abundance; then reactions that linked the amino acids together to form chains; and finally reactions that permitted the chains of amino acids to assume the configurations that they had in native proteins. During the war years it was established beyond reasonable doubt that DNA (deoxyribonucleic acid) was the material that determined the hereditary characters of the cell.[2] DNA turned out to be a particularly attractive substance to have been the primaeval biological molecule, for it was capable of self-replication and, once formed, would have been able to generate copies of itself. DNA was also a chain of linked subunits, so the precursors of DNA were subjected to experiments similar in principle to those that had been done with the precursors of proteins. Then attention turned to RNA (ribonucleic acid), the linear transcript of DNA. RNA proved to be even more attractive than DNA, for it was found that RNA could not only replicate itself; it could also direct the synthesis of DNA and could act as a catalyst for other chemical reactions. All these plausible reconstructions of past events are, and perhaps can only be, speculative; but they do not seem to me to be more speculative than explanations of the origin of life that invoke divine intervention.

Notes

1. A. I. Oparin (1953), *The origin of life* (2nd edn) (trans. S. Morgulis). Dover, N.Y.
2. O. Avery, C. MacLeod, and M. McCarty (1944), *J. Exptl. Med.* **79**: 137.

Index